AQA Chemistry Revision Guide

A LEVEL

Ted Lister
Janet Renshaw
Emma Poole

OXFORD
UNIVERSITY PRESS

Great Clarendon Street, Oxford, OX2 6DP, United Kingdom

Oxford University Press is a department of the University of Oxford.
It furthers the University's objective of excellence in research,
scholarship, and education by publishing worldwide. Oxford is a
registered trade mark of Oxford University Press in the UK and in
certain other countries

© Ted Lister, Janet Renshaw, and Emma Poole 2017

The moral rights of the authors have been asserted

First published in 2017

All rights reserved. No part of this publication may be reproduced,
stored in a retrieval system, or transmitted, in any form or by any
means, without the prior permission in writing of Oxford University
Press, or as expressly permitted by law, by licence or under terms agreed
with the appropriate reprographics rights organization. Enquiries
concerning reproduction outside the scope of the above should be sent
to the Rights Department, Oxford University Press,
at the address above.

You must not circulate this work in any other form and you must
impose this same condition on any acquirer

British Library Cataloguing in Publication Data
Data available

978-0-19-835184-9

3

Paper used in the production of this book is a natural, recyclable
product made from wood grown in sustainable forests.
The manufacturing process conforms to the environmental regulations
of the country of origin.

Printed and bound by CPI Group (UK) Ltd, Croydon, CR0 4YY

Cover: Omer N Raja / Shutterstock

Artwork by Q2A Media

AS/A Level course structure

This book has been written to support students studying for AQA A Level Chemistry. It covers all A Level modules from the specification. The sections covered are shown in the contents list, which also shows you the page numbers for the main topics within each section. If you are studying for AS Chemistry, you will only need to know the content in the blue box.

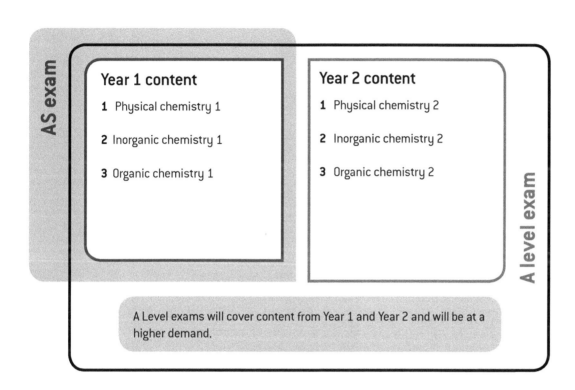

How to use this book vi

Section 1 Physical chemistry 1

Chapter 1 Atomic structure 2
1.1 Fundamental particles 2
1.2 Mass number, atomic number, and isotopes 2
1.3 The arrangement of the electrons 3
1.4 The mass spectrometer 4
1.5 Electron arrangements and ionisation energy 5
Practice questions 6

Chapter 2 Amount of substance
2.1 Relative atomic and molecular masses, the Avogadro constant, and the mole 7
2.2 Moles in solution 8
2.3 The ideal gas equation 9
2.4 Empirical and molecular formulae 10
2.5 Balanced equations and related calculations 10
2.6 Balanced equations, atom economies, and percentage yields 11
Practice questions 13

Chapter 3 Bonding
3.1 The nature of ionic bonding 14
3.2 Covalent bonding 15
3.3 Metallic bonding 16
3.4 Electronegativity – bond polarity in covalent bonds 17
3.5 Forces acting between molecules 19
3.6 The shapes of molecules and ions 20
3.7 Bonding and physical properties 21
Practice questions 23

Chapter 4 Energetics 24
4.1 Endothermic and exothermic reactions 24
4.2 Enthalpy 24
4.3 Measuring enthalpy changes 25
4.4 Hess's law 27
4.5 Enthalpy changes of combustion 28
4.6 Representing thermochemical cycles 29
4.7 Bond enthalpies 30
Practice questions 31

Chapter 5 Kinetics 32
5.1 Collision theory 32
5.2 Maxwell–Boltzmann distribution 32
5.3 Catalysts 34
Practice questions 35

Chapter 6 Equilibria 36
6.1 The idea of equilibrium 36
6.2 Changing the conditions of an equilibrium reaction 36
6.3 Equilibrium reactions in industry 37
6.4 The equilibrium constant, K_c 38
6.5 Calculations using equilibrium constant expressions 39
6.6 The effect of changing conditions on equilibria 40
Practice questions 41

Chapter 7 Oxidation, reduction, and redox reactions 42
7.1 Oxidation and reduction 42
7.2 Oxidation states 42
7.3 Redox equations 44
Practice questions 45

Section 2 Inorganic chemistry 1

Chapter 8 Periodicity 46
8.1 The Periodic Table 46
8.2 Trends in the properties of elements of Period 3 46
8.3 More trends in the properties of elements of Period 3 48
8.4 A closer look at ionisation energies 48
Practice questions 49

Chapter 9 Group 2, the Alkaline Earth metals 50
9.1 The physical and chemical properties of Group 2 50
Practice questions 52

Chapter 10 Group 7(17), the halogens 53
10.1 The halogens 53
10.2 The chemical reactions of halogens 54
10.3 Reactions of halide ions 55
10.4 Uses of chlorine 55
Practice questions 56

Section 3 Organic chemistry 1

Chapter 11 Introduction to organic chemistry 57
11.1 Carbon compounds 57
11.2 Nomenclature – naming organic compounds 59
11.3 Isomerism 60
Practice questions 62

Chapter 12 Alkanes 63
12.1 Alkanes 63
12.2 Fractional distillation of crude oil 65
12.3 Industrial cracking 66
12.4 Combustion of alkanes 68
12.5 The formation of halogenoalkanes 70
Practice questions 71

Chapter 13 Halogenoalkanes 72
13.1 Halogenoalkanes – introduction 72
13.2 Nucleophilic substitution in halogenoalkanes 73
13.3 Elimination reactions in halogenoalkanes 74
Practice questions 76

Chapter 14 Alkenes 77
14.1 Alkenes 77
14.2 Reactions of alkenes 78
14.3 Addition polymers 79
Practice questions 80

Chapter 15 Alcohols 81
15.1 Alcohols – an introduction 81
15.2 Ethanol production 82
15.3 The reactions of alcohols 83
Practice questions 85

Chapter 16 Organic analysis 86
16.1 Test-tube reactions 86
16.2 Mass spectroscopy 87
16.3 Infrared spectroscopy 88
Practice questions 89

Section 1 Physical chemistry 2

Chapter 17 Thermodynamics — 90
- 17.1 Enthalpy change — 90
- 17.2 Born–Haber cycles — 90
- 17.3 More enthalpy changes — 93
- 17.4 Why do chemical reactions take place? — 94
- Practice questions — 96

Chapter 18 Kinetics — 97
- 18.1 The rate of chemical reactions — 97
- 18.2 The rate expression and order of reaction — 98
- 18.3 Determining the rate equation — 99
- 18.4 The Arrhenius equation — 100
- 18.5 The rate-determining step — 101
- Practice questions — 102

Chapter 19 Equilibrium constant K_p — 103
- 19.1 Equilibrium constant, K_p for homogeneous systems — 103

Chapter 20 Electrode potentials and electrochemical cells — 104
- 20.1 Electrode potentials and the electrochemical series — 104
- 20.2 Predicting the direction of redox reactions — 106
- 20.3 Electrochemical cells — 108
- Practice questions (19 and 20) — 111

Chapter 21 Acids, bases, and buffers — 112
- 21.1 Defining an acid — 112
- 21.2 The pH scale — 113
- 21.3 Weak acids and bases — 114
- 21.4 Acid–base titrations — 115
- 21.5 Choice of indicators for titrations — 116
- 21.6 Buffer solutions — 117
- Practice questions — 119

Section 2 Inorganic chemistry 2

Chapter 22 Periodicity — 120
- 22.1 Reactions of Period 3 elements — 120
- 22.2 The oxides of elements in Period 3 — 121
- 22.3 The acidic/basic nature of the Period 3 oxides — 123
- Practice questions — 124

Chapter 23 The transition metals — 125
- 23.1 The general properties of transition metals — 125
- 23.2 Complex formation and the shape of complex ions — 126
- 23.3 Coloured ions — 129
- 23.4 Variable oxidation states of transition elements — 130
- 23.5 Catalysis — 132
- Practice questions — 133

Chapter 24 Reactions of inorganic compounds in aqueous solutions — 134
- 24.1 The acid–base chemistry of aqueous transition metal ions — 134
- 24.2 Ligand substitution reactions — 135
- 24.3 A summary of acid–base and substitution reactions of some metal ions — 135
- Practice questions — 136

Section 3 Organic chemistry 2

Chapter 25 Nomenclature and isomerisms — 137
- 25.1 Naming organic compounds — 137
- 25.2 Optical isomerism — 138
- 25.3 Synthesis of optically active compounds — 139
- Practice questions — 140

Chapter 26 Compounds containing the carbonyl group — 141
- 26.1 Introduction to aldehydes and ketones — 141
- 26.2 Reactions of the carbonyl group in aldehydes and ketones — 142
- 26.3 Carboxylic acids and esters — 143
- 26.4 Reactions of carboxylic acids and esters — 143
- 26.5 Acylation — 145
- Practice questions — 147

Chapter 27 Aromatic chemistry — 148
- 27.1 Introduction to arenes — 148
- 27.2 Arenes – physical properties, naming, and reactivity — 149
- 27.3 Reactions of arenes — 149
- Practice questions — 151

Chapter 28 Amines — 152
- 28.1 Introduction to amines — 152
- 28.2 The properties of amines as bases — 153
- 28.3 Amines as nucleophiles and their synthesis — 154

Chapter 29 Polymerisation — 156
- 29.1 Condensation polymers — 156
- Practice questions (28 and 29) — 158

Chapter 30 Amino acids, proteins, and DNA — 159
- 30.1 Introduction to amino acids — 159
- 30.2 Peptides, polypeptides, and proteins — 159
- 30.3 Enzymes — 161
- 30.4 DNA — 162
- 30.5 The action of anti-cancer drugs — 164
- Practice questions — 165

Chapter 31 Organic synthesis and analysis — 166
- 31.1 Synthetic routes — 166
- 31.2 Organic analysis — 167
- Practice questions — 168

Chapter 32 Structure determination — 169
- 32.1 Nuclear magnetic resonance (NMR) spectroscopy — 169
- 32.2 Proton NMR — 171
- 32.3 Interpreting proton, 1H, NMR spectra — 172
- Practice questions — 174

Chapter 33 Chromatography — 175
- 33.1 Chromatography — 175
- Practice questions — 177

Synoptic questions — 178
Answers to practice questions — 184
Answers to summary questions — 194
Answers to synoptic questions — 207
Data — 210
Periodic table — 213

Specification references
→ At the beginning of each topic, there are specification references to allow you to monitor your progress.

This book contains many different features. Each feature is designed to foster and stimulate your interest in chemistry, as well as supporting and developing the skills you will need for your examinations.

Worked example
Step-by-step worked solutions.

Key term
Pulls out key terms for quick reference.

Summary questions

1 These are short questions at the end of each topic.

2 They test your understanding of the topic and allow you to apply the knowledge and skills you have acquired.

3 The questions are ramped in order of difficulty. Lower-demand questions have a paler background, with the higher-demand questions having a darker background. Try to attempt every question you can, to help you achieve your best in the exams.

Synoptic link
These highlight how the sections relate to each other. Linking different areas of chemistry together becomes increasingly important, and you will need to be able to do this.

Shows you what you need to know for the practical aspects of the exam.

Revision tip
Prompts to help you with your revision.

Maths skill
A focus on maths skills.

Common misconception
Common student misunderstandings clarified.

 ### Go further
To push you a little further.

Question and model answers
Sample answers to exam-style questions.

Chapter 32 Practice questions

1 The ^{1}H NMR spectrum of ethanoic acid, CH_3COOH, is shown below.

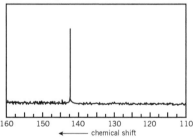

  a How many different types of hydrogen atom are present in a molecule of ethanoic acid? *(1 mark)*

  b Explain why there is no spin–spin splitting of the two peaks. *(1 mark)*

  c What peak is normally found on NMR spectra at δ = 0. What is the purpose of this peak? *(2 marks)*

  d Suggest a solvent for running the spectrum of ethanoic acid and explain your choice. *(2 marks)*

2 a How many peaks would you expect in the ^{13}C NMR spectrum of pentan-3-one? Explain your answer. *(2 marks)*

  b How many peaks would you expect in the ^{1}H NMR spectrum of pentan-3-one? Explain your answer. *(2 marks)*

  c Predict the ratio of peak heights in the ^{1}H NMR spectrum of pentan-3-one. *(1 mark)*

  d Predict how the peaks in the ^{1}H NMR spectrum of pentan-3-one will be split. Explain your answer. *(4 marks)*

  e Name the rule that predicts spin–spin splitting of ^{1}H NMR peaks. *(1 mark)*

3 The ^{13}C NMR spectrum of the compound buckminsterfullerene, C_{60}, is shown.

  a The spectrum has one line only. What does this confirm about the environment of each carbon atom in C_{60}? *(1 mark)*

  b What compound is used in NMR spectroscopy to calibrate the zero of the spectrum? State its name (or abbreviation) and its formula. *(2 marks)*

  c What are the units of chemical shift used in the spectrum? *(1 mark)*

  d How many peaks would you expect to see in the ^{1}H NMR spectrum of buckminsterfullerene? Explain your answer. *(2 marks)*

pentan-3-one

buckminsterfullerene, C_{60}

1.1 Fundamental particles
1.2 Mass number, atomic number, and isotopes
Specification reference: 3.1.1

> **Key term**
>
> **Atomic number (Z):** The number of protons in the nucleus.

> **Key term**
>
> **Mass number (A):** The number of protons and neutrons in the nucleus.

> **Key term**
>
> **Isotope:** The isotopes of an atom of an element contain different numbers of neutrons but the same number of protons and also of electrons. So they have the same atomic number but a different mass number.

Summary questions

1. Write down the numbers of protons, neutrons, and electrons in the following:
    a. $^{23}_{11}Na$
    b. $^{16}_{8}O^{2-}$
    c. $^{80}_{35}Br$
    d. $^{40}_{19}K^+$ (4 marks)

2. Explain why chlorine-35 and chlorine-37 are chemically identical.   (1 mark)

3. Describe the relative mass and relative charge of the sub-atomic particles found in the nucleus of an atom.   (2 marks)

Atoms and sub-atomic particles

Atoms are made from sub-atomic or fundamental particles called protons, neutrons, and electrons. The masses and charges of all these fundamental particles are very small, so we always look at them relative to the mass and charge of a proton.

Particle	Relative mass	Relative charge
proton	1	+1
neutron	1	0
electron	$\frac{1}{1836}$	−1

The mass of an electron is so small it is often considered to be negligible. Protons and neutrons are located in the nucleus of the atom while electrons orbit the nucleus.

Atomic number and mass number

Ensure that you can recall the definitions of atomic number and mass number.

- Atoms can be represented: $^A_Z X$
- Number of protons = atomic number (Z)
- Number of electrons = atomic number (Z) (ONLY for neutral atoms)
- Number of neutrons = mass number − atomic number (A − Z)

For ions (particles that have lost or gained electrons) the charge needs to be taken into account.

Beryllium atoms $^9_4 Be$ protons = 4, electrons = 4, neutrons = 9 − 4 = 5

Fluoride ions $^{19}_9 F^-$ protons = 9, electrons = 9 + 1 = 10, neutrons = 19 − 9 = 10

Calcium ions $^{40}_{20} Ca^{2+}$ protons = 20, electrons = 20 − 2 = 18, neutrons = 40 − 20 = 20

Isotopes

The chemistry of an atom is determined by how its electrons behave. All the isotopes of an element have the same number of electrons and therefore the same electron arrangement. This means all the isotopes of an element will react chemically in an identical way. However the difference in the number of neutrons between isotopes may cause slight variations in physical properties such as boiling point. Isotopes of the same element will be chemically identical.

There are two isotopes of chlorine

$^{37}_{17} Cl$ protons = 17, electrons = 17, neutrons = 20

$^{35}_{17} Cl$ protons = 17, electrons = 17, neutrons = 18

1.3 The arrangement of electrons
Specification reference: 3.1.1

Nature of electrons

Over time, chemists' understanding of electrons has improved, as technology has got better. Chemists today believe that electrons have some properties of particles, some of waves, and at other times behave as clouds of charge.

Energy levels

Electrons are in constant motion around the nucleus of an atom.

- Electrons are found in energy levels or shells.
- These are split into sub-levels or sub-shells with different maximum numbers of electrons.
- Different types of sub-level contain different numbers of orbitals.
- Each orbital can hold two electrons (one spinning up and one spinning down).
- The shape of an orbital tells you where an electron is most likely to be found.

Synoptic link
This will be useful when you study electron arrangements in Topic 1.5, Electron arrangements and ionisation energy.

Key term
Orbital: An orbital is a region where up to two electrons can exist.

Revision tip
The 2p sub-level or sub-shell can contain up to six electrons but a 2p orbital still only contains up to two electrons.

▼ **Table 1** *Deducing the maximum number of electrons in each energy level*

Energy level	1	2		3			4			
Type of sub-level	s	s	p	s	p	d	s	p	d	f
Number of orbitals in sub-level	1	1	3	1	3	5	1	3	5	7
Maximum number of electrons in sub-level	2	2	6	2	6	10	2	6	10	14
Maximum number of electrons in level	2	8		18			32			

Electron configuration

▼ **Table 2** *Deducing spin diagrams*

Element	Electron configuration	Spin diagrams										
		1s	2s	2p			3s	3p				
H	$1s^1$	↑										
He	$1s^2$	↑↓										
Li	$1s^2 2s^1$	↑↓	↑									
C	$1s^2 2s^2 2p^2$	↑↓	↑↓	↑	↑							
O	$1s^2 2s^2 2p^4$	↑↓	↑↓	↑↓	↑	↑						
P	$1s^2 2s^2 2p^6 3s^2 3p^3$	↑↓	↑↓	↑↓	↑↓	↑↓	↑↓	↑	↑	↑		

- The period that an element is in determines its highest energy level.
- The group that an element is in determines its outer electron configuration.

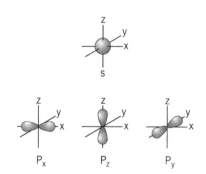

▲ **Figure 1** *Shapes of orbitals*

Summary questions

1. Sketch the shape of an s-orbital. *(1 mark)*

2. Write the full electron configuration for:
   a Na   b S   c Cl   d Be   *(4 marks)*

3. Write full electron configurations for:
   a magnesium, Mg^{2+} ion b sulfide, S^{2-} ion
   c oxide, O^{2-} ion
   d sodium, N^+ ion   *(4 marks)*

1.4 The mass spectrometer
Specification reference: 3.1.1

Synoptic link
Mass spectrometry can also be used to measure relative molecular masses and much more, as you will see in Topic 16.2, Mass spectrometry.

Mass spectrometer
The mass spectrometer is a very sensitive machine used to analyse samples of elements in terms of the isotopes they contain and their relative amounts and also to discover information about the structure of organic molecules. You must learn names and brief explanations for each step in the electro spray ionisation time of flight (TOF) mass spectrometer.

- **Vacuum** The whole apparatus is kept under vacuum so that no air is present. This stops the ions formed in the apparatus from colliding with molecules in the air.
- **Ionisation** The sample is dissolved in a volatile solvent and then forced through an electrically charged thin hollow needle to produce a stream of positively charged droplets. The solvent dissolves leaving single positively charged ions.
- **Acceleration** The positive ions accelerate towards the negatively charged plate.
- **Ion drift** The positive ions pass through a hole in the negatively charged plate. They form a beam that travels through the flight tube.
- **Detection** The positively charged ions reach the detector and are recorded. Lighter ions travel more quickly so take less time to reach the detector.
- **Data analysis** The information from the detector is sent to a computer which produces a mass spectrum.

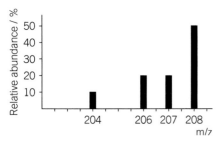

▲ **Figure 1** *The mass spectrum of lead*

Relative atomic mass $\quad A_r = \dfrac{\text{mean mass of an atom of the element}}{\text{mass of one atom of }^{12}\text{C}} \times 12$

Relative molecular mass $\quad M_r = \dfrac{\text{mean mass of a molecule}}{\text{mass of one atom of }^{12}\text{C}} \times 12$

Relative isotopic mass $\quad = \dfrac{\text{mass of one atom of the isotope}}{\text{mass of one atom of }^{12}\text{C}} \times 12$

Synoptic link
You will find out more about relative atomic mass in Topic 2.1, Relative atomic and molecular masses, the Avogadro constant, and the mole.

Relative atomic mass, A_r
The mass of all atoms is measured relative to the mass of a ^{12}C atom. Relative atomic mass can be calculated using the data from a mass spectrum. From the mass spectrum of lead opposite we obtain the following data:

m/z	204	206	207	208
Relative abundance / %	10.0	20.0	20.0	50.0

- m/z is the same as the mass of the ion if the charge is +1.
- Relative abundance tells us the proportion of each isotope present in the sample.
- The A_r of lead can then be calculated as a weighted mean average.

$$A_r = \dfrac{(204 \times 10.0) + (206 \times 20.0) + (207 \times 20.0) + (208 \times 50.0)}{100} = 207$$

When answering calculation questions it is essential to record your answer to an appropriate number of significant figures. The data in the question is given to three significant figures so the final answer must also be recorded to three significant figures.

Summary questions

1. Name the six steps involved in electro spray ionisation time of flight mass spectrometer. *(1 mark)*

2. Use the spectral data of neon given below to determine its relative atomic mass. Give your answer to an appropriate number of significant figures.

m/z	Relative abundance (%)
20.0	90.9
21.0	0.300
22.0	8.80

(2 marks)

1.5 Electron arrangements and ionisation energy

Specification reference: 3.1.1

Ionisation energy

The values obtained for ionisation energies provide substantial evidence for the existence of energy levels and sub-levels.

First ionisation energies of Group 2 elements

Down the group, the nuclear charge (number of protons in the nucleus) increases. However, the electrons in the outer shell are lost more easily. This is because down the group the electrons in the outer shell are further from the nucleus and there is more shielding. As a result the first ionisation energy decreases as you go down a group.

Successive ionisation energies of magnesium

Successive ionisation energies provide further evidence for the existence of energy levels.

- There is a general increase in the energy needed to remove each electron from magnesium.
- This is because the electron is being removed from an ion with an increasing positive charge.
- There is a very big increase between the 2nd and 3rd ionisation energies this is because the 3rd electron is being removed from an electron shell closer to the nucleus.
- The group to which an element belongs can be determined by identifying where the big jump in ionisation energy occurs.

First ionisation energies of the elements in Period 3

The graph of first ionisation energies for Period 3 elements provides substantial support for the existence of energy sub-levels.

- The number of protons increases across period 3 so there is an increase in the charge on the nucleus. As a result the force of attraction between the nucleus and the outer electron increases.
- The number of electrons also increases but these go into the same energy level so are at a similar distance from the nucleus and experience similar shielding. As a result there is an increase in first ionisation energy across Period 3.

Drops in ionisation energy occur at two points on the graph.

- Between magnesium and aluminium there is a decrease in ionisation energy because the outer electron in aluminium is in a p sub-level which is of slightly higher energy than the s sub-level, so is easier to remove.
  Mg $1s^2\ 2s^2\ 2p^6\ 3s^2$ Al $1s^2\ 2s^2\ 2p^6\ 3s^2\ 3p^1$
- Between phosphorus and sulfur there is a slight decrease because in sulfur two of the p-electrons are paired and this pair repel each other and the outer electron is easier to remove.
  P $1s^2\ 2s^2\ 2p^6\ 3s^2\ 3p^3$ S $1s^2\ 2s^2\ 2p^6\ 3s^2\ 3p^4$

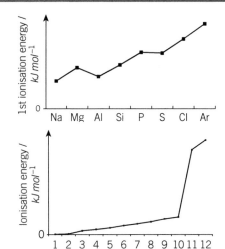

▲ **Figure 1** (a) Successive ionisation energies of magnesium (b) First ionisation energies of the Period 3 elements

Key term

Ionisation energy: First ionisation energy is the energy required to remove 1 electron from each atom in 1 mole of gaseous atoms forming 1 mole of ions with a single positive charge.

e.g. The equation for the first ionisation (of magnesium) is:

$$Mg(g) \rightarrow Mg^+(g) + e^-$$

Summary questions

1. Write down the equation that represents the third ionisation energy of magnesium, Mg. *(2 marks)*

2. Explain why the first ionisation energies of Group 2 elements decrease down the group. *(2 marks)*

3. Sketch a graph showing the successive ionisation energies of oxygen. Label the graph in detail, explaining the patterns it shows. *(2 marks)*

Chapter 1 Practice questions

1. Use the Periodic Table to deduce the full electron configuration of: *(2 marks)*
   a Mg
   b Na⁺

2. A sample of magnesium was analysed and found to contain three isotopes. The percentage abundance of each isotope is shown below.

Isotope	Percentage abundance / %
^{24}Mg	78.0
^{25}Mg	10.0
^{26}Mg	12.0

   Use the information in the table to calculate the relative atomic mass of this sample of magnesium. Give your answer to three significant figures. *(2 marks)*

3. Complete the table to show the relative charge and relative mass of these sub-atomic particles. *(3 marks)*

Sub-atomic particle	Relative charge	Relative mass
proton		
neutron		
electron		

4. A non-metallic element can be recognised from its relative atomic mass. Analysis of a sample of the non-metallic element revealed the following percentage abundances.

Isotope	% abundance	Relative isotopic mass
1	25.0	37.0
2	75.0	35.0

   a Define the term isotope. *(1 mark)*
   b Name the analytic method used to determine the percentage abundance of the non-metallic element. *(1 mark)*
   c Calculate the relative atomic mass of the non-metallic element. Give your answer to two significant figures. *(2 marks)*
   d Use your answer and the data sheet to suggest the identity of the non-metallic element. *(1 mark)*

5. Which of these atoms has the smallest atomic radius?
   Ar
   P
   Al
   Na *(1 mark)*

6. Consider the table below. Which line shows the correct number of protons, neutrons, and electrons in the ion?

	Ion	Protons	Neutrons	Electrons
A	^{23}Na⁺	11	12	11
B	^{19}F⁻	9	9	8
C	^{16}O^{2-}	8	8	8
D	^{27}Al^{3+}	13	14	10

   (1 mark)

2.1 Relative atomic and molecular masses, the Avogadro constant, and the mole

Specification reference: 3.1.2

Calculations using relative molecular mass and relative formula mass

a Calculate the relative molecular mass of oxygen, O_2.
$$M_r = 16.0 \times 2 = 32.0$$

b Calculate the relative molecular mass of carbon dioxide, CO_2.
$$M_r = 12.0 + (16.0 \times 2) = 44.0$$

c Calculate the relative formula mass of sodium chloride, NaCl.
$$M_r = 23.0 + 35.5 = 58.5$$

d Calculate the relative formula mass of calcium hydroxide, $Ca(OH)_2$.
$$M_r = 40.1 + (17.0 \times 2) = 74.1$$

The mole and the Avogadro constant

Chemists are interested in how many atoms, molecules, and ions take part in reactions. All of these particles are very small so you cannot determine their mass. Instead you carry out calculations using the concept of the mole.

- The mol is the unit for amount of substance.
- One mole of a substance contains the same number of particles as there are atoms in exactly 12 g of ^{12}C.
- The number of particles in one mole of a substance is the Avogadro number (after Amedeo Avogadro), 6.022×10^{23}.

Amount of substance

The amount of a substance is readily calculated from the relative atomic mass or the relative molecular mass of a substance using the relationship: mass = $M_r \times n$. You must be able to manipulate this relationship in order to calculate mass, M_r, or n given suitable data. n is the number of moles of the substance.

Calculations using mass = $M_r \times n$

a Calculate the mass in grams of 2 moles of Mg atoms.
$$\text{mass} = A_r \times n = 24.3 \times 2 = 48.6 \text{ g}$$

b Calculate the number of moles of Na atoms in 6.5 g of Na atoms.
$$n = \frac{\text{mass}}{A_r} = \frac{6.5}{23.1} = 0.28$$

c Calculate the number of moles of MgO in 2×10^{-6} g of MgO.
$$n = \frac{\text{mass}}{M_r} = \frac{2 \times 10^{-6}}{40.3} = 4.96 \times 10^{-8}$$

d Calculate the M_r of aluminium oxide, Al_2O_3, if 0.5 moles of aluminium oxide has a mass of 51 g
$$M_r = \frac{\text{mass}}{n} = \frac{51}{0.5} = 102$$

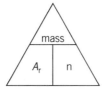 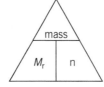

A_r = relative atomic mass
n = number of moles

M_r = relative molecular mass / formula mass
n = number of moles

▲ **Figure 1** *Amount of substance*

Key term

Relative atomic mass: The relative atomic mass, A_r, is the weighted average mass of an atom of an element, taking into account its naturally occurring isotopes, relative to $\frac{1}{12}$ of the relative atomic mass of an atom of carbon-12. The relative atomic mass of an element is displayed on the Period Table.

Key term

Relative molecular mass: The relative molecular mass, M_r, of a molecule is the mass of that molecule compared to $\frac{1}{12}$ of the relative atomic mass of an atom of carbon-12. The relative molecular mass of a substance is found by adding together the relative atomic mass of all the atoms in the molecule.

Key term

Relative formula mass: The term relative formula mass is used for ionic compounds as ionic compounds do not contain molecules. Relative formula mass also has the symbol M_r. It is calculated using the same method used for relative molecular mass. Relative atomic mass, relative molecular mass, and relative formula mass do not have units.

Summary questions

1 Calculate the following. Remember to show all your working. Give your answers to an appropriate number of significant figures.
   a The number of moles of CaO in 5.61 g of CaO
   b The mass of 0.150 moles of K
   c The mass of 0.32 moles of LiOH   *(3 marks)*

2.2 Moles in solution

Specification reference: 3.1.2

> **Revision tip**
> Make sure you give your answer to the required number of decimal places.

Concentrations

- Concentration is the amount of a substance in moles dissolved in $1\,dm^3$ of solution ($1\,dm^3 = 1000\,cm^3$). Concentration has units of $mol\,dm^{-3}$.
- This means that a $2.00\,mol\,dm^{-3}$ solution of sulfuric acid contains 2.00 moles of sulfuric acid dissolved in $1.00\,dm^3$ of water. As the M_r of sulfuric acid is 98.1.
- This is $2.00 \times 98.1 = 196.2\,g$ of sulfuric acid dissolved in $1\,dm^3$ of solution.

The relationship between concentration and number of moles is: $n = c \times v$
n = number of moles, c = concentration in dm^3, v = volume of solution in dm^3
Note – The volume used in the concentration expression is needed in dm^3. To convert dm^3 to cm^3 you must multiply by 1000. To convert cm^3 to dm^3 you must divide by 1000.

Concentration calculations

a Calculate the number of moles of hydrochloric acid in $50\,cm^3$ of $0.1\,mol\,dm^{-3}$ solution.

Number of moles = $\frac{50}{1000} \times 0.1 = 0.005\,mol$

b Calculate the volume, in cm^3, of $0.0500\,mol\,dm^{-3}$ sodium hydroxide solution that would contain 1.15×10^{-3} moles.

Volume = $1.15 \times 10^{-3} \times \frac{1000}{0.0500} = 23.0\,cm^3$

c $50.0\,cm^3$ of nitric acid contains 1.25×10^{-3} moles. Calculate the concentration of the acid.

Concentration = $1.25 \times 10^{-3} \times \frac{1000}{50.0} = 0.250\,mol\,dm^{-3}$

Acid–base titration calculations

Follow these systematic steps to carry out a titration calculation.

$25.0\,cm^3$ of sodium hydroxide solution is exactly neutralised by $21.4\,cm^3$ of $0.500\,mol\,dm^{-3}$ hydrochloric acid. What was the concentration of the sodium hydroxide solution?

Step 1	Write a balanced equation $NaOH(aq) + HCl(aq) \rightarrow NaCl(aq) + H_2O(l)$
Step 2	Write out the data given in the question under the equation. $NaOH(aq) + HCl(aq) \rightarrow NaCl(aq) + H_2O(l)$ $25.0\,cm^3$ $\;21.4\,cm^3$ $\;\;\;?\;\;\;0.500\,mol\,dm^{-3}$
Step 3	Convert the substance you know the most about into moles. number of moles of $HCl(aq)$: $n = c \times v = 0.500 \times (21.4 \div 1000) = 0.0107$
Step 4	Use the balanced equation to determine the moles of the unknown substance. 1 mol HCl reacts with 1 mol NaOH. So number of moles of NaOH = 0.0107
Step 5	Convert the number into the units asked for in the question. concentration of sodium hydroxide solution: $c = \frac{n}{v} = \frac{0.0107}{(25.0 \div 1000)}$ $= 0.428\,mol\,dm^{-3}$

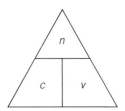

n = number of moles
c = concentration in $mol\,dm^{-3}$
v = volume of solution in dm^3

▲ **Figure 1** *Concentration is the amount of a substance in moles dissolved in $1\,dm^3$ of solution*

> **Summary questions**
>
> 1 1.00 moles of a nitric acid, HNO_3, is dissolved in $500\,cm^3$ of water. What is the concentration of the nitric acid? *(1 mark)*
>
> 2 $25.0\,cm^3$ of NaOH(aq) is exactly neutralised by $27.6\,cm^3$ of $0.150\,mol\,dm^{-3}$ $HNO_3(aq)$. Calculate the concentration of the NaOH(aq) solution. Give your answer to an appropriate number of significant figures. *(3 marks)*

2.3 The ideal gas equation
Specification reference: 3.1.2

The ideal gas equation
Finding the pressure of a gas
What is the pressure exerted by 0.200 moles of chlorine gas in a vessel of 6.00 m³ at a temperature of 298 K?

$$p = \frac{nRT}{V} = \frac{0.200 \times 8.31 \times 298}{6.00} = 82.5 \text{ Pa}$$

Finding the volume of a gas
What is the volume of 6.40×10^{-4} moles of hydrogen gas at a pressure of 1.00 Pa at 273 K?

$$V = \frac{nRT}{P} = \frac{6.40 \times 10^{-4} \times 8.31 \times 273}{1.00} = 1.45 \text{ m}^3$$

Calculating the volume of gas released in a reaction
2.00 g of calcium carbonate is reacted with an excess of 0.200 mol dm⁻³ hydrochloric acid. Calculate the volume of carbon dioxide (in m³) released at standard temperature and pressure.

$$CaCO_3(s) + 2HCl(aq) \rightarrow CaCl_2(aq) + H_2O(l) + CO_2(g)$$

Step 1: Calculate the number of moles of calcium carbonate.

$$n = \frac{\text{mass}}{M_r} = \frac{2.00}{(40.1 + 12.0 + (3 \times 16.0))} = \frac{2.00}{100.1} = 0.01998$$

Step 2: Use the balanced equation to determine the number of moles of carbon dioxide.

$$\text{ratio } CaCO_3 : CO_2 \quad 1:1$$

$$n \text{ } CO_2 = 0.01998$$

Step 3: Use the ideal gas equation to calculate the volume of carbon dioxide.

$$V = \frac{nRT}{p} = \frac{0.01998 \times 8.31 \times 298}{1.00 \times 10^5} = 4.95 \times 10^{-4} \text{ m}^3$$

Summary questions

1. Nitrogen and hydrogen react in the Haber process to form ammonia. Calculate the volume of ammonia that will be formed when 90.0 cm³ of hydrogen reacts completely with nitrogen if the temperature and pressure are kept constant. $N_2(g) + 3H_2(g) \rightarrow 2NH_3(g)$ *(1 mark)*

2. Calculate the volume occupied by 2.31 moles of hydrogen gas at a pressure of 2×10^5 Pa and a temperature of 32°C. *(1 mark)*

3. Magnesium metal reacts with hydrochloric acid to form hydrogen gas.
$$Mg(s) + 2HCl(aq) \rightarrow MgCl_2(aq) + H_2(g)$$
Calculate the volume of gas released when 0.20 g of magnesium reacts with an excess of acid at 298 K and 1×10^5 Pa. *(3 marks)*

Revision tip
$pV = nRT$ p: pressure in pascals (Pa)
V: volume in m³
n: amount of a gas in moles
R: gas constant = 8.31 J K⁻¹ mol⁻¹
T: temperature in K

Key term
The ideal gas equation: The ideal gas equation links the volume of a gas to moles, pressure, and temperature of the gas.

Revision tip
The ideal gas equation is always stated in terms of pressure.

Converting quantities
Temperatures
0 °C = 273 K
To convert from °C to K add 273
To convert from K to °C subtract 273

Volumes
1 cm³ = 10⁻⁶ m³
To convert from cm³ to m³ multiply by 10⁻⁶
To convert from m³ to cm³ divide by 10⁻⁶

Pressures
1 kPa = 10³ Pa
To convert from kPa to Pa multiply by 10³
To convert from Pa to kPa divide by 10³

Revision tip
1.00 mol of any gas takes up a volume of 24 dm³ at RTP.

2.4 Empirical and molecular formula
2.5 Balanced equations and related calculations

Specification reference: 3.1.2

Revision tip
The empirical formula of a compound can be calculated if the percentage composition of the compound is known.

Element	Ca	C	O
% by mass	40.06	11.99	47.95
÷ A_r	0.999	0.999	2.997
÷ smallest	1	1	3

Synoptic link
Once we know the formula of a compound we can use techniques such as infrared spectroscopy and mass spectrometry to help work out its structure, see Chapter 16, Organic analysis.

Key terms
Empirical formula: The empirical formula of a compound is the simplest whole number ratio of the atoms of the elements in a compound.

Molecular formula: The molecular formula is the actual number of atoms of each element in one molecule of the compound. The molecular formula can be the same as the empirical formula or a whole number multiple of the empirical formula.

Revision tip
Make sure you include the full formula at the end of the calculation.

Calculating empirical formula
A compound contains 40.06% calcium, 11.99% carbon, and 47.95% oxygen by mass. Determine its empirical formula. Show all your working.

Empirical formula: $CaCO_3$

Calculating the molecular formula
A compound has an empirical formula of CH_2 and a relative molecular mass of 98.0. Calculate the molecular formula of the compound.

The relative mass of the empirical formula = 12.0 + (2 × 1.0) = 14.0

$$\frac{\text{the relative formula mass}}{\text{relative mass of the empirical formula}} = \frac{98.0}{14.0} = 7$$

The molecular formula = C_7H_{14}

Writing a balanced equation
Magnesium reacts with oxygen to form magnesium oxide.

Step 1: Work out the identities of the reactants and products.

$$\text{magnesium} + \text{oxygen} \rightarrow \text{magnesium oxide}$$

Step 2: Construct formulae for each of the reactants and products.

$$Mg + O_2 \rightarrow MgO$$

Step 3: Balance the equation so that the number of each type of atom is the same on each side of the equation. Work from left to right.

$$2Mg + O_2 \rightarrow 2MgO$$

Step 4: Add state symbols using (s) for solid, (l) for liquid, (g) for gas and (aq) for aqueous, a solution in water.

$$2Mg(s) + O_2(g) \rightarrow 2MgO(s)$$

Summary questions

1. Write the balanced symbol equation for the reaction between sodium and chlorine. Include state symbols. *(2 marks)*

2. Calculate the empirical formula of an oxide of iron which contains 69.9% iron by mass. *(3 marks)*

3. A compound is analysed and found to have an empirical formula of CH_2 and a relative molecular mass of 42.0. Deduce the molecular formula of the compound. *(2 marks)*

2.6 Balanced equations, atom economies, and percentage yields
Specification reference: 3.1.2

Balanced chemical equations in unfamiliar situations

You will be expected to write and balance simple equations for reactions that you have studied in the unit. You may also be given an unfamiliar equation and be asked to balance it.

> **Key term**
>
> **Atom economy:**
> % atom economy = $\dfrac{\text{mass of desired product}}{\text{total mass of all products}} \times 100$

Balancing an equation for an unfamiliar reaction

Ammonia, NH_3, reacts with sodium to form sodium amide, $NaNH_2$, and hydrogen. Write a balanced equation for this reaction.

Step 1: Write formulae for the reactants on the left of an equation and for the products on the right.

$$NH_3 + Na \rightarrow NaNH_2 + H_2$$

Step 2: Work through the equation from left to right balancing each atom in turn. There is one nitrogen atom on the left and one on the right so this equation balances in terms of nitrogen.

There are three hydrogen atoms on the left and four on the right. In order to provide enough hydrogen atoms for the right you must place a two in front of the ammonia.

$$2NH_3 + Na \rightarrow NaNH_2 + H_2$$

You now need to check the nitrogen atoms again; there are two on the left so you need to place a two in front of the sodium amide.

$$2NH_3 + Na \rightarrow 2NaNH_2 + H_2$$

The nitrogen and hydrogen atoms are now balanced. To finish the equation you need to place a two in front of the sodium atom.

$$2NH_3 + 2Na \rightarrow 2NaNH_2 + H_2$$

Atom economy

The atom economy of a chemical reaction is the proportion of reactants that are converted into useful products.

- Processes with high atom economy are more efficient and produce less waste.
- This is important for sustainable development.

Atom economy is calculated by looking at the mass of desired product in relation to the total mass of the products. These may be expressed as mass in grams or as relative molecular mass. The total mass of the reactants is equal to the total mass of the products.

Calculating atom economy

a A chemical reaction produces 48 g of a desired product and 27 g of undesired product. Calculate the percentage atom economy.

$$\% \text{ atom economy} = \dfrac{\text{mass of desired product}}{\text{total mass of all products}} \times 100$$

$$= \dfrac{48}{75} \times 100 = 64\%$$

11

Amount of substance

> **Maths skill**
> Calculate the relative mass of the desired product and the relative mass of all the products first.

b The reaction of methane with steam produces hydrogen gas along with carbon monoxide as a waste product. Calculate the percentage atom economy in relation to hydrogen gas.

$$CH_4(g) + H_2O(g) \rightarrow 3H_2(g) + CO(g)$$

$$\% \text{ atom economy} = \frac{\text{relative mass of desired product}}{\text{total relative mass of all products}} \times 100$$

$$= \frac{3 \times 2}{((3 \times 2) + 28)} \times 100 = \frac{6}{34} \times 100 = 17.6\%$$

Note that the ratios of each reacting species have been taken into account here.

Percentage yield

The actual yield of a product shown as a percentage of the expected yield.

$$\text{percentage yield} = \frac{\text{actual yield}}{\text{theoretical yield}} \times 100$$

> **Worked example**
>
> Q A reaction has a theoretical yield of 1.1 moles of product.
>
> A student carries out the reaction and produces 0.60 moles of product. What is the percentage yield of this reaction?
>
> A Percentage yield $= \frac{0.60}{1.1} \times 100 = 55\%$

> **Summary questions**
>
> 1 Ethene, C_2H_4, reacts with hydrogen bromide, HBr, to form bromoethane, C_2H_5Br. Calculate the percentage yield for this reaction if 2.00 g of ethene forms 5.80 g of bromoethane. *(3 marks)*
>
> 2 Chlorine gas can be obtained from the electrolysis of brine. The equation for this process is:
> $2NaCl(aq) + 2H_2O(l) \rightarrow 2NaOH(aq) + Cl_2(g) + H_2(g)$.
> Calculate the atom economy for producing chlorine. *(2 marks)*
>
> 3 The Haber process for making ammonia by reacting nitrogen and hydrogen gases
> $N_2(g) + 3H_2(g) \rightleftharpoons 2NH_3(g)$
> typically has a percentage yield of around 15%. Explain what is meant by the term percentage yield and compare this with the atom economy for this process. *(2 marks)*

Chapter 2 Practice questions

1. A 10.0 g sample of metal was analysed and found to contain 20.0% iron by mass.
    a. Calculate the amount, in mol, of iron in the sample. Give your answer to an appropriate number of significant figures. *(1 mark)*
    b. Calculate the number of atoms of iron in the sample of metal.
       $N_A = 6.02 \times 10^{23}\,\text{mol}^{-1}$ *(1 mark)*

2. A chloride of phosphorus was analysed and found to contain 14.9% phosphorus by mass.
    a. Define the term *empirical formula*. *(1 mark)*
    b. Calculate the empirical formula of the compound. *(2 marks)*

3. 25.0 cm³ of a solution of sodium hydroxide of concentration 0.100 mol dm⁻³ was reacted with a dilute solution of hydrochloric acid of unknown concentration.
   23.5 cm³ of the hydrochloric acid was required for complete neutralisation.
   The equation for the reaction is shown below.

   $HCl(aq) + NaOH(aq) \rightarrow NaOH(aq) + H_2O(l)$

    a. Calculate the amount, in mol, of sodium hydroxide in the 25.0 cm³ sample. *(2 marks)*
    b. Calculate the amount, in mol, of hydrochloric acid required. *(1 mark)*
    c. Calculate the concentration of the hydrochloric acid solution. *(2 marks)*

4. A student placed 1.50 g of zinc metal into a beaker containing 50 cm³ of 0.2 mol dm⁻³ hydrochloric acid.
   The equation for the reaction is shown below.

   $Zn(s) + 2HCl(aq) \rightarrow ZnCl_2(aq) + H_2(g)$

    a. What would you see during the reaction? *(2 marks)*
    b. Calculate the amount of zinc added. *(2 marks)*
    c. Calculate the amount of hydrochloric acid in the beaker. *(2 marks)*
    d. Would there be any zinc left at the end of the experiment? Explain your answer. *(1 mark)*

5. How many atoms are present in 5.00 g of calcium carbonate, $CaCO_3$?
   $N_A = 6.02 \times 10^{23}\,\text{mol}^{-1}$ *(1 mark)*

   A 3.01×10^{22}
   B 1.50×10^{23}
   C 0.0500
   D 9.03×10^{22}

6. What is the amount, in mol, in 50.0 cm³ of a 0.200 mol dm⁻³ solution of copper(II) sulfate? *(1 mark)*

   A 10 mol
   B 0.0100 mol
   C 0.0250 mol
   D 0.002 mol

7. Ethanol, C_2H_5OH, can be produced by reacting ethene, C_2H_4, with steam, H_2O. The equation for the reaction is shown below.

   $C_2H_4(g) + H_2O(g) \rightarrow C_2H_5OH(g)$

   Deduce the atom economy of this reaction.
   A 50%   B 100%   C 46%   D 61%   *(1 mark)*

3.1 The nature of ionic bonding

Specification reference: 3.1.3

> **Key term**
>
> **Ionic bond:** An electrostatic attraction between ions of opposite charge.

> **Revision tip**
>
> Covalent bonding occurs between non-metal atoms.
>
> Ionic bonding occurs between metal and non-metal atoms.
>
> Metallic bonding occurs between metal atoms.
>
> Make sure you talk about the right sort of bonding in your answers.

Formation of ions

All chemical bonds are forces of attraction. Ionic bonds occur when a metal and a non-metal react to form a compound.

- Metal atoms lose electrons forming cations. For example, a lithium atom loses 1 electron to form a lithium ion.

    $Li \rightarrow Li^+ + e^-$ The lithium ion has electron configuration $1s^2$.

- Non-metal atoms gain electrons forming anions. For example, a fluorine atom gains 1 electron to form a fluoride ion.

    $F + e^- \rightarrow F^-$ The fluoride ion has electron configuration $1s^2\ 2s^2\ 2p^6$.

Ionic lattices

Ionic compounds are made up of lattice structures.

- In an ionic compound each ion is surrounded by ions of opposite charge.
- For example, in the sodium chloride lattice, each chloride ion is surrounded by six sodium ions and vice versa.
- This structure is repeated throughout the ionic compound.
- As a result ionic compounds are said to have giant structures.

Constructing ionic formulae

Ionic formulae show the ratio of cations to anions present in the ionic lattice. A formula for an ionic compound is constructed from the formulae of the cations and anions present. In most cases you can work out the charges of the cation and anion using the Periodic Table.

- For metal atoms the charge on the ion is the same as the group number of the metal.
- Ionic compounds have a neutral charge overall so the number of positive charges must match the number of negative charges. For example: Barium chloride contains Ba^{2+} ions and Cl^- ions, so has the formula $BaCl_2$.
- Some ionic compounds contain compound ions or molecular ions.

For example:

- Sodium sulfate contains Na^+ ions and SO_4^{2-} ions, so has the formula Na_2SO_4. Aluminium sulfate contains Al^{3+} ions and SO_4^{2-} ions, so has the formula $Al_2(SO_4)_3$.

Properties of ionic compounds

- Ionic compounds have giant ionic structures. They always have high melting points and are solid at room temperature. Ionic compounds conduct electricity when molten or dissolved as the ions can move. However, they do not conduct electricity when solid as the ions cannot move. Ionic compounds are brittle and shatter easily when hit.

> **Summary questions**
>
> 1 Give the formula of
>   a sodium bromide
>   b calcium hydroxide
>   c magnesium chloride.
>   *(3 marks)*
>
> 2 Why does sodium chloride conduct when molten but not when solid? *(1 mark)*

3.2 Covalent bonding
Specification reference: 3.1.3

Sharing of electrons

Non-metal atoms can achieve full outer shells either by accepting electrons from metal atoms or by sharing pairs of electrons.

- Covalent bonds can form between identical atoms or different atoms.
- As with ionic bonding only the outer energy level electrons are involved in bonding.
- A pair of electrons is shared between two atoms.
- The bond is held together by the attraction between each nucleus involved in the bond and the pair of electrons.

Covalent bonds can be represented by dot-and-cross diagrams. They can also be shown as a straight line between the atoms. This is called a displayed formula.

Multiple bonds

Non-metal atoms can form double or triple covalent bonds by sharing more than one pair of electrons. These are represented in molecular formulae by multiple lines between the atoms. Double and triple bonds are stronger than single bonds.

Bonding with lone pairs

Lone pairs of electrons are able to form dative covalent bonds with atoms that have vacant orbitals. Dative covalent bonds are shown in displayed formulae by an arrow.

For example:

- The ammonium ion, NH_4^+, has a dative covalent bond between the nitrogen atom and one of the hydrogen atoms.
- In carbon monoxide, CO, the oxygen atom forms a double covalent bond and also a co-ordinate bond.
- In the NH_3BH_3 molecule the boron atom has a vacant orbital and accepts a pair of electrons from the nitrogen atom.

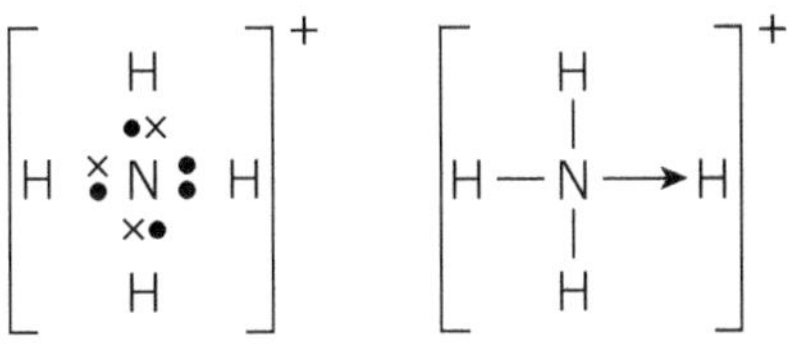

▲ **Figure 1** *Examples of molecules containing dative covalent bonds*

Simple molecular substances

Simple molecular substances can be solid, liquid, or gas at room temperature but they tend to have relatively low melting points. There are strong bonds within the molecules but only weak forces of attraction between molecules. Simple molecules are poor electrical conductors. Some simple molecules dissolve in water.

Key terms

Single covalent bond: A shared pair of electrons.

Double covalent bond: Two shared pairs of electrons.

Triple covalent bond: Three shared pairs of electrons.

Key terms

Lone pair of electrons: A pair of electrons not involved in bonding.

Dative covalent (or co-ordinate) bond: A shared pair of electrons in which one of the atoms contributes both electrons.

Summary questions

1. What is a covalent bond? *(2 marks)*

2. What is a dative covalent bond? *(2 marks)*

3. Why is nitrogen, N_2, a gas at room temperature? *(2 marks)*

3.3 Metallic bonding
Specification reference: 3.1.3

Key term

Metallic bond: The electrostatic force of attraction between metal ions and the delocalised electrons in a metallic lattice.

The metallic bond

Atoms of metals are held together by metallic bonds. In a metallic bond:

- Each metal atom forms a positive ion (cation).
- The positive ions are arranged into a regular lattice structure.
- The ions in the structure are very close to each other so the electrons that are lost when the metal atoms form ions become delocalised.
- Delocalised electrons are not attracted to any particular ion.
  - As a result, these electrons are free to move through the metal.

For example, magnesium atoms have the electron configuration $1s^2\ 2s^2\ 2p^6\ 3s^2$. The 3s electrons are delocalised from each atom forming magnesium ions of charge 2+. These have the electron configuration $1s^2\ 2s^2\ 2p^6$.

The metallic bond is the strong attraction between the positive ions in the lattice and the delocalised sea of electrons.

Metallic bond strength

Metallic bonds do not all have the same strength. If they did, then all metals would melt and boil at the same temperature.

The strength of a metallic bond is dependent on:

- the charge of the ions in the lattice
- the number of electrons in the sea of delocalised electrons.

The boiling point graph for sodium, magnesium, and aluminium shows this. Magnesium needs more energy to boil than sodium as the attraction between the Mg^{2+} ions in the lattice and the sea of electrons is greater than between the Na^+ and the sea of electrons. The force of attraction is even stronger in aluminium, which has Al^{3+} ions.

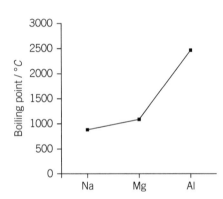

▲ **Figure 1** *Boiling points of sodium, magnesium, and aluminium*

Properties of metals

- Metals are good conductors of heat and electricity.
- Metals have high melting points and boiling points.
- Metals are malleable and ductile as the layers of ions can slip over each other.

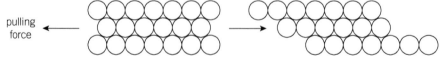

▲ **Figure 2** *Metals are malleable and ductile*

Summary questions

1. Draw a simple sketch of the structure of a piece of aluminium. Label your sketch to show how aluminium is able to conduct electricity. *(2 marks)*

2. List the following metals in order of increasing melting point. Explain the order you have chosen: sodium, aluminium, magnesium. *(2 marks)*

3. Consider the term *ductile*. Explain why metals are ductile. *(1 mark)*

3.4 Electronegativity – bond polarity in covalent bonds

Specification reference: 3.1.3

Charge distribution in covalent molecules

The covalent bond is held together by the attraction between the nuclei of the two atoms involved in the bond and the pairs of electrons.

- If the bond is between identical atoms this sharing is equal, e.g. H–H, Cl–Cl
- If the atoms are different then the sharing may be unequal, e.g. H–F. The fluorine atom is much better at attracting the pair of electrons than hydrogen.

Electronegativity

If the two atoms in a bond have different electronegativites then the more electronegative element has a greater share of the electrons. Non-metal elements at the top of Groups 5, 6, and 7 such as N, O, F, and Cl are the most electronegative.

Polar bonds

Polar bonds are those in which the pair of electrons is not shared equally because there is a large difference in electronegativity between the atoms.

- The more electronegative element has a partial negative charge, shown by $\delta-$.
- The less electronegative element has a partial positive charge, shown by $\delta+$.

In the hydrogen chloride molecule:

- The chlorine atom is more electronegative so has a partial negative charge.
- The hydrogen atom therefore has a partial positive charge.
  - As a result, The H–Cl *molecule* can be described as polar.

In the carbon dioxide molecule:

- The oxygen atoms are more electronegative so have a partial negative charge.
- The carbon atom has two partial positive charges as it is bonded to two oxygen atoms. As a result, The C=O *bond* can be described as polar.

H 2.2							He –
Li 1.0	Be 1.6	B 2.0	C 2.5	N 3.0	O 3.4	F 4.0	Ne –
Na 0.9	Mg 1.3	Al 1.6	Si 1.9	P 2.2	S 2.6	Cl 3.2	Ar –
K 0.8	Ca 1.0					Br 3.0	Kr 3.0
Rb 0.8						I 2.7	Xe 2.6

▲ **Figure 3**

> **Key term**
>
> **Covalent bonding:** Covalent bonding is the sharing of one or more pairs of electrons.

> **Key term**
>
> **Electronegativity:** Electronegativity is the ability of an atom to withdraw electron density from a covalent bond.

$$\overset{\delta+}{H}-\overset{\delta-}{Cl}$$

▲ **Figure 1** *Hydrogen chloride*

$$\overset{\delta-}{O}=\overset{2\delta+}{C}=\overset{\delta-}{O}$$

▲ **Figure 2** *Carbon dioxide*

> **Key terms**
>
> **Polar bond:** A covalent bond between atoms with different electronegativities.
>
> **Dipole:** Opposite charges separated by a short distance in a molecule or ion.

Bonding

Polar bonds and polar molecules

A molecule that contains polar bonds may not be a polar molecule. To find out if a molecule is polar:

- Draw the molecule (in three dimensions if necessary, remembering about the influence of lone pairs).
- Label any polar bonds using the $\delta+$, $\delta-$ convention.
- Then examine the shape of the molecule.
- If the molecule is symmetrical the polar bonds cancel out and the molecule is not polar.
- If the molecule is not symmetrical polar bonds do not cancel out and the molecule is polar.

Worked example:

Are carbon dioxide and water polar molecules?

$\overset{\delta-}{O}=\overset{\delta+}{C}=\overset{\delta-}{O}$ Carbon dioxide has two polar bonds but there is no net dipole as the bonds are symmetrical. Carbon dioxide is a non-polar molecule.

Water has two polar bonds and has a net dipole beacause the polar bonds do not cancel each other out. Water is polar molecule. This has a big influence on the properties of water.

Summary questions

1. Label the following bonds to indicate their polarity.
   a H–F   b C–Cl   c O–N   d S=O
   (4 marks)

2. Draw the following molecules. Label any polar bonds and state whether the molecule is polar.
   a hydrogen bromide
   b hydrogen sulfide
   c ammonia
   d fluorine oxide, F_2O
   (8 marks)

3. Arrange the substances below in order of their covalent character starting with the least covalent:
   a $NaCl$, $AlCl_3$, $MgCl_2$ b NaI, $NaCl$, $NaBr$
   (2 marks)

3.5 Forces acting between molecules
Specification reference: 3.1.3

Van der Waals' forces

A van der Waals' force is a force of attraction between a temporary dipole on one molecule and an induced dipole on another molecule.

Temporary dipoles

Electrons in a molecule are constantly moving.

- This means that the electron cloud around an atom or within a non-polar molecule is not static.
- At any instant in time the distribution of the electrons may be uneven although on average they are distributed evenly. As a result, a non-polar molecule may have a temporary dipole.

Induced dipoles

The presence of a temporary dipole in one atom or molecule can cause a dipole to form in a nearby atom or molecule. This dipole is called an induced dipole. The induced dipole can then induce a dipole in a neighbouring atom or molecule. The net effect of this is a force of attraction between the particles called a temporary dipole–induced dipole force.

The more electrons a molecule has the stronger the van der Waals' forces between molecules will be and the higher its boiling temperature is.

Permanent dipole–dipole forces

Molecules with a permanent dipole have regions of different electron density within them. These molecules are described as being polar. It is easy to see that if two polar molecules come near each other in space there will be an attraction between them. For the hydrogen chloride molecule in Figure 1, the electronegative chlorine of one HCl molecule will attract the electropositive hydrogen of another.

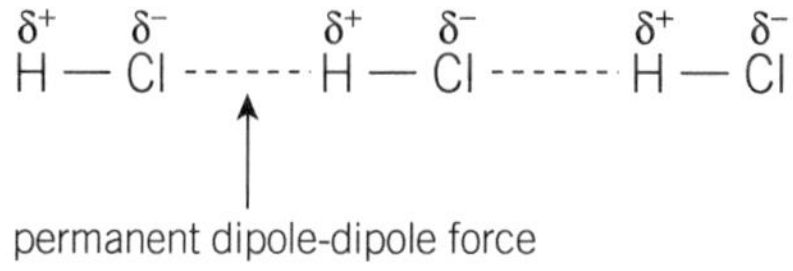

Hydrogen bonding

Hydrogen bonds are an especially strong permanent dipole–dipole force, which exist between molecules that contain very electronegative elements.

For a hydrogen bond to occur:

- The molecule must have an O, N, or F atom bonded to a hydrogen atom.
- The molecule to which it is attracted must contain an O, N, or F atom.

There are, of course, many other molecules, for example, methanol that are capable of hydrogen bonding.

> **Revision tip**
> Intermolecular forces are weak attractions that exist between molecules. There are three types of intermolecular force; van der Waals', permanent dipole-dipole interactions, and hydrogen bonds.

$\delta^+ \quad \delta^- \qquad \delta^+ \quad \delta^- \qquad \delta^+ \quad \delta^-$
H — Cl ------ H — Cl ------ H — Cl

permanent dipole-dipole force

▲ **Figure 1** *Hydrogen chloride*

> **Revision tip**
> When drawing a diagram to show hydrogen bonding you must include the following:
> - labelled dipoles on every molecule
> - lone pairs on the O, N, or F atom
> - a dotted line to represent the hydrogen bond
> - hydrogen bonds drawn at an angle of 180°

Summary questions

1. Draw a labelled diagram to show the hydrogen bonding in
   a hydrogen fluoride
   b ammonia *(2 marks)*

2. Methane, CH_4 and ethane, C_2H_4 are hydrocarbons. Predict which of these hydrocarbons has the highest melting point. Explain your answer. *(2 marks)*

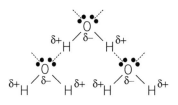

Hydrogen bonding in water

▲ **Figure 2** *Hydrogen bonding in water*

3.6 The shapes of molecules and ions
Specification reference: 3.1.3

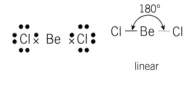

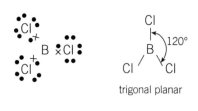

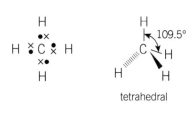

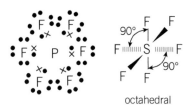

▲ Figure 1 The shapes of molecules

Valence shell electron pair repulsion theory

The shapes of molecules and ions are determined by applying valence shell electron pair repulsion theory:

- Valence shell electrons are those in the outermost energy level.
- These pairs of electrons repel each other.
- The shape of a molecule is such that the distance between the pairs of electrons is as large as possible.
- Multiple bonds have the same repelling effect as single bonds.
- Lone pairs of electrons are more repelling than bonding pairs of electrons. As a result the bond angles are slightly smaller when lone pairs are present.

▼ Table 1 Predicting the orbital shape and bond angle of molecules

Pairs of electrons	Basic shape	Bond angles
2	linear	180°
3	trigonal planar	120°
4	tetrahedral	109.5°
6	octahedral	90°

Molecules containing one or more lone pairs of electrons
Tetrahedral

Both ammonia, NH_3, and water, H_2O, are molecules containing four pairs of electrons overall.

- In ammonia one of these pairs of electrons is a lone pair. As a result ammonia has a pyramidal shape with a bond angle of 107°
- In water two of the pairs of electrons are lone pairs. As a result water has a bent (or non-linear) shape with a bond angle of 104.5°.
- The repulsive effect of a lone pair of electrons is greater than a bond pair. As a result the bond angles in ammonia and water are slightly smaller than that of methane.

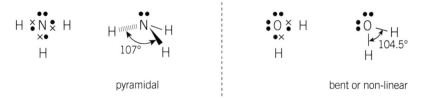

▲ Figure 2 Ammonia and water molecules

Revision tip
The shapes of molecular ions can be worked out in the same way as for uncharged molecules.

- Remember to take account of the charge on the ion when working out the dot-and-cross diagram.
- Positively charged ions have fewer electrons than the original atom.
- Negatively charged ions have gained extra electrons.

Summary questions

1 Draw a dot-and-cross diagram for each of the following molecules. Then use it to draw a labelled diagram of the molecule showing all bond angles. Include the name of the shape of each molecule.
 a BF_3 b CCl_4 c PH_3 d H_2S *(4 marks)*

3.7 Bonding and physical properties
Specification reference: 3.1.3

Changing state 🧪

When a solid is melted or a liquid frozen, a liquid boiled or a gas condensed it is said to have changed state.

- All changes of state involve changes in energy.
- When a solid melts or a liquid boils (or vaporises), energy is used to break the forces between the atoms, molecules, or ions involved.
- As the change of state occurs, the temperature stays constant because the energy provided to the system is used to break the forces of attraction between the particles.

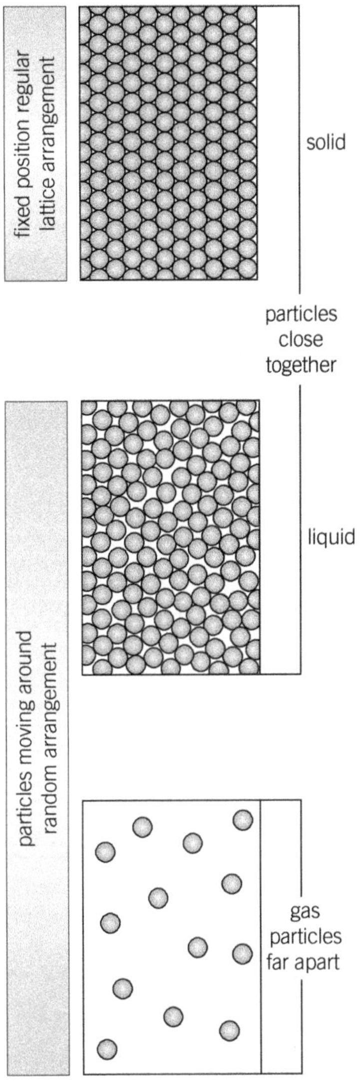

Metals

When melting a pure metal the attraction between the lattice of positive ions and the delocalised sea of electrons is broken. This is a strong force so requires a large amount of energy, which means that metals have high melting points. The melting points increase as the charge on the metal ion and the number of delocalised electrons increases.

Ionic substances

Remember that an ionic bond is an electrostatic attraction between ions of opposite charge. When an ionic substance melts, the energy provided is used to break this attraction. The attraction is strong so ionic substances are hard to melt and are solid at room temperature.

Giant molecular substances

Giant molecular substances have a large network of covalent bonds. Melting these substances involves a large amount of energy as lots of strong covalent bonds must be broken in order to change state. You must be able to explain why diamond and graphite have such high melting points.

Simple molecular substances

Melting simple molecular substances requires the breaking of the intermolecular force between the molecules.

- These forces are much weaker than the covalent bond that exists between the atoms within the molecule.

You must be able to identify the type of force between molecules and relate this to changes of state.

▲ **Figure 1** *States of matter*

> **Synoptic link**
> You will learn more about enthalpy in Topic 4.2, Enthalpy.

> **Revision tip**
> The energy required to change state from a solid to a liquid is called the enthalpy change of fusion while the energy required to change state from a liquid to a gas is called the enthalpy change of vaporisation.

> **Revision tip**
> Simple molecular structures have strong covalent bonds within the molecules but only weak intermolecular forces of attraction between molecules.

Bonding

Revision tip
Metals conduct because the delocalised electrons can move (not the ions).

▼ **Table 1** *Intermolecular forces*

Name of force	Relative strength of force
permanent dipole–dipole attraction	stronger than van der Waals' force, weaker than hydrogen bond
hydrogen bond	the strongest intermolecular force
van der Waals' force	weakest intermolecular force

Conducting electricity

In order to conduct electricity, charged particles must be able to move through a substance when a voltage is applied:

- In a metal the delocalised electrons are free to move.
- In a molten ionic substance the ions are free to move. Note that ionic solids do not conduct electricity as the ions are held in a fixed position but they do conduct if the solid is dissolved in another substance so that the ions can move.
- Simple molecular substances and giant molecular substances do not conduct electricity as there are no charged particles. The exception to this is carbon – graphite is able to conduct as there are delocalised electrons between the layers of carbon atoms.

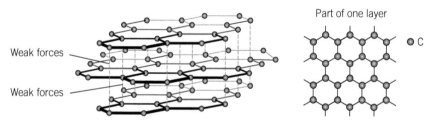

▲ **Figure 2** *The structure of graphite*

Summary questions

1. Complete the table below showing the type of bonding in each substance

Name of substance	Formula	Type of structure	Type of bonding
magnesium	Mg	giant ionic	
sodium chloride			ionic
chlorine			
graphite			

(1 mark)

2. Arrange the following substances in order of melting point starting with the lowest. Explain the order you have chosen.
H_2O, $MgCl_2$, Cl_2. *(2 marks)*

Chapter 3 Practice questions

1 The shape of methane, CH_4 and ammonia, NH_3, molecules can be predicted using valence electron pair repulsion theory.

   a Methane, CH_4, has a bond angle of 109.5°. Name the shape of the methane molecule and explain why methane has the bond angle of 109.5°. *(3 marks)*

   b State and explain the bond and angle in an ammonia, NH_3, molecule. *(3 marks)*

2 Define the term electronegativity. *(2 marks)*

3 a Describe the bonding and structure of magnesium. Include a diagram in your answer. *(4 marks)*

   b Explain why magnesium is a good electrical conductor. *(1 mark)*

4 Water has a higher melting point and boiling point than expected compared with the other Group 6 hydrides. Explain why water has a higher melting point than expected. Include a diagram in your answer. *(4 marks)*

5 a Complete the table below to show the structure, type of bonds, and electrical conductivity of magnesium and magnesium oxide, MgO. *(3 marks)*

Substance	Mg	MgO
Structure		
Type of bonding		
Electrical conductivity when solid		

   b Explain why magnesium oxide conducts electricity when it is dissolved in water? *(1 mark)*

   c Explain why magnesium oxide has a high melting point. *(2 marks)*

6 Strontium oxide is an ionic compound.

   a Define the term *ionic bond*. *(2 marks)*

   b Draw a dot-and-cross diagram to show the bonding in strontium oxide. Draw the outer shell of electrons only. *(1 mark)*

   c Explain why strontium oxide has a high melting point. *(3 marks)*

7 Which one of these substances contains ionic bonds?

   A H_2O

   B Na_2O

   C CO_2

   D Mg *(1 mark)*

8 Which of these equations shows the most polar carbon-halogen bond?

   A C–Cl

   B C–F

   C C–I

   D C–Br *(1 mark)*

4.1 Endothermic and exothermic reactions
4.2 Enthalpy

Specification reference: 3.1.4

Exothermic and endothermic reactions

Energy must be taken in to break existing bonds (an endothermic process), whilst energy is given out when new bonds are formed (an exothermic process). During chemical reactions there is often a difference between the amount of energy taken in to break the existing bonds and the amount of energy given out when the new bonds are formed.

Exothermic reactions

- The substances involved in exothermic reactions get hotter because chemical energy is being changed into thermal (heat) energy.
- During the reaction the chemicals lose energy. The energy that is lost by the chemicals is gained by the surroundings. As a result the enthalpy change (ΔH) for an exothermic reaction is always negative.

Endothermic reactions

Some reactions take in more energy to break existing bonds then they release when new bonds are formed. These reactions are described as endothermic.

- The substances involved in endothermic reactions get colder because thermal energy is taken in from the surroundings.
- The energy that is lost by the surroundings is gained by the chemicals. As a result the enthalpy change (ΔH) for an endothermic reaction is always positive.

Standard enthalpy change of formation

- The standard enthalpy of formation, $\Delta_f H^\ominus$, is the enthalpy change when one mole of a compound is formed from its elements in their standard states, under standard conditions.

  As a result, the standard enthalpy change of formation of an element in its standard state is zero.

Standard enthalpy change of combustion

The standard enthalpy of combustion, $\Delta_c H^\ominus$, is the enthalpy change when one mole of a compound is completely burnt in oxygen under standard conditions.

The reaction involving $\Delta_c H^\ominus$ of methane, CH_4, is represented as

$$CH_4(g) + 2O_2(g) \rightarrow CO_2(g) + 2H_2O(g)$$

$$\Delta_c H^\ominus \text{ at } 25°C = -890.3 \text{ kJ mol}^{-1}$$

> **Key terms**
>
> **Enthalpy:** The heat energy stored in a chemical system.
>
> **Exothermic reactions:** Chemicals lose energy to the surrounding. The ΔH for exothermic reactions is negative.

> **Key term**
>
> **Endothermic reactions:** Chemicals gain energy from the surrounding. The ΔH for exothermic reactions is positive.

> **Key term**
>
> **Enthalpy change (ΔH):** The heat energy change measured under conditions of constant pressure.

> **Summary questions**
>
> 1. What is $\Delta_f H^\ominus$ of $N_2(g)$? *(1 mark)*
>
> 2. What is the equation that represents the standard enthalpy change of combustion of ethanol? Include state symbols. *(2 marks)*

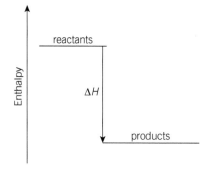

▲ **Figure 1** An exothermic reaction

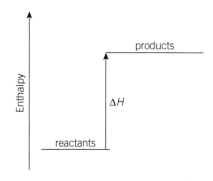

▲ **Figure 2** An endothermic reaction

4.3 Measuring enthalpy changes
Specification reference: 3.1.4

Calorimetry

Worked example

Q A student added an excess of magnesium powder to a solution of 0.100 mol dm^{-3} hydrochloric acid. The student used 50.0 cm^3 of hydrochloric acid. The temperature at the start of the experiment was 21.0 °K. The student measured a maximum temperature of 31.0 °K. Assume that the specific heat capacity of the solution is 4.2 J g^{-1} °K^{-1} and that the solution has a density of 1.00 g cm^{-3}. Calculate the heat energy change of this reaction.

A $q = mc\Delta T$

$q = 50.0 \times 4.2 \times (31.0 - 21.0) = 2100$ J

Notice how the calculated results can only be given to 2 significant figures as the least accurate measurement (specific heat capacity of the solution) was only given to 2 significant figures.

Calculating the enthalpy change of a reaction

The enthalpy change of a reaction is calculated as the heat energy change per mole and has units of kJ mol^{-1}.

To calculate the enthalpy change of a reaction:

Calculate the heat energy change involved.

Divide the heat energy change for the reaction by 1000.

Calculate the number of moles.

Then divide the heat energy change by the number of moles involved.

Remember to include the sign for the enthalpy change.

Exothermic reactions release energy to the surrounding and have a negative sign.

Endothermic reactions take in energy from the surroundings and have a positive sign.

Revision tip
Calculating the heat change of a reaction

The heat energy change for a reaction (q, measured in joules) is given by the equation

$q = mc\Delta T$

Where m = the mass of the surroundings (g), c = the specific heat capacity of the surroundings, (J g^{-1} K^{-1}) and ΔT is the temperature change (final temperature − initial temperature), (K).

This gives a value for the heat change of the reaction in J.

To give an answer in kilojoules remember to divide by 1000. Normally it is acceptable to give an answer either as joules or kilojoules so it is essential to include the correct unit with your answer.

Worked example

Q A chemist placed 25.0 cm^3 of 0.200 mol dm^{-3} sodium hydroxide solution into a polystyrene cup then added 25.0 cm^3 of 0.200 mol dm^{-3} hydrochloric acid solution. The initial temperature of both solutions was 22.5 °C. The final temperature of the solutions was 29.5 °C. Assume that the specific heat capacity of the solution is 4.18 J g^{-1} K^{-1} and that all the solutions have a density of 1.00 g cm^{-3}.

Calculate the enthalpy change of the reaction.

A $q = mc\Delta T$

$= 50 \times 4.18 \times 7.0 = 1463$ J

$= 1.463$ kJ

The number of moles of both reactants is

$\frac{25.0}{1000} \times 0.200 = 0.00500$ mol

Energetics

> The enthalpy change = 1.463 kJ/0.00500 mol
>
> $$= -293 \text{ kJ mol}^{-1}$$
>
> The final answer is given to three significant figures and the minus sign added to show the reaction is exothermic.

Measuring enthalpy changes more accurately

A major source of error with this experiment is heat loss. Polystyrene is a good insulator and helps to prevent heat being lost to the surrounding during exothermic reactions or gained from the surrounding by endothermic reactions. Adding a lid to the polystyrene cup would help prevent heat loss but would make it hard to read the thermometer accurately. However adding more insulation to the polystyrene cup would improve the experimental results.

Adding the reactant chemicals together quickly, using powders (which have a high surface area) and stirring the chemicals all ensure that the chemicals react quickly and allow a maximum temperature change to be recorded before the product chemicals start to cool down to (or warm up to) room temperature.

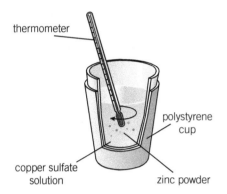

▲ **Figure 1** *This simple calorimeter can be used to calculate the heat energy change during the displacement reaction between zinc and copper sulfate*

Common misconception

In exothermic reactions heat energy is given out so the temperature of the surroundings increases.

But ΔH of the system is negative.

Summary questions

1. What is the major source of error in a calorimetry experiment carried out in a beaker? How could this error be reduced? *(2 marks)*

2. A student placed 50.0 cm³ of 1.00 mol dm⁻³ copper sulfate solution in a polystyrene cup. The solution has an initial temperature of 21.0 °C. They then quickly added an excess of zinc powder and recorded the maximum temperature as 30.5 °C. Assume that the specific heat capacity of the solution is 4.18 J g⁻¹ K⁻¹ and that all the solutions have a density of 1.00 g cm⁻³
   a. Why was zinc powder used?
   b. Calculate the heat energy change of this reaction.
   c. Calculate the enthalpy change of the reaction. *(5 marks)*

4.4 Hess's law
Specification reference: 3.1.4

Hess's law

The enthalpy change of a chemical reaction is independent of the route by which the reaction is achieved and depends only on the initial and final states.

- As a result you can use Hess's law to find enthalpy changes for reactions that cannot be measured directly in the laboratory.
- The enthalpy change from the reactants A to the products B is the same as the sum of the enthalpy changes from A to C and from C to B.
- We can use $\Delta_f H^\ominus$ and $\Delta_c H^\ominus$ values.

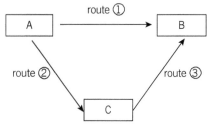

▲ Figure 1 route 1 = route 2 + route 3

Using enthalpies of formation

The standard enthalpy of formation ($\Delta_f H^\ominus$) is the enthalpy change when one mole of a compound is formed from its elements in their standard states under standard conditions.

Standard enthalpies of formation can be used to calculate enthalpy changes that cannot be measured directly.

- Notice that the arrow is drawn from the elements to the chemicals.

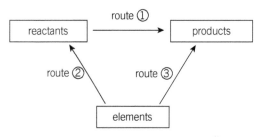

▲ Figure 2 Using standard enthalpy of formation values: route 1 = – route 2 + route 3

Worked example

Q Calculate the enthalpy change for the combustion of ethene, $C_2H_4(g)$.

$C_2H_4(g) + 3O_2(g) \rightarrow 2CO_2(g) + 2H_2O(g)$

Given that

$\Delta_f H^\ominus C_2H_4(g) = +52\,kJ\,mol^{-1}$

$\Delta_f H^\ominus CO_2(g) = -394\,kJ\,mol^{-1}$

$\Delta_f H^\ominus H_2O(g) = -242\,kJ\,mol^{-1}$

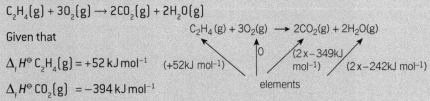

▲ Figure 3 The Hess's law diagram for the combustion of ethene

A The unknown enthalpy change $\Delta_r H^\ominus$

$= -(+52\,kJ\,mol^{-1}) + (2 \times -394\,kJ\,mol^{-1}) + (2 \times -242\,kJ\,mol^{-1})$

$= -1324\,kJ\,mol^{-1}$

Q Calculate the enthalpy change for the reaction below.

$CH_4(g) + H_2O(g) \rightarrow CO(g) + 3H_2(g)$

Given that

$\Delta_f H^\ominus CH_4(g) = -75\,kJ\,mol^{-1}$

$\Delta_f H^\ominus CO(g) = -110\,kJ\,mol^{-1}$

$\Delta_f H^\ominus H_2O(g) = -242\,kJ\,mol^{-1}$

▲ Figure 4 The Hess's law diagram for the reaction

A The unknown enthalpy change

$= -(-75\,mol^{-1}) - (-242\,kJ\,mol^{-1}) + (-110\,kJ\,mol^{-1})$

$= +207\,kJ\,mol^{-1}$

Summary questions

1. What is Hess's law? *(1 mark)*

2. In $\Delta_f H^\ominus$ what are the conditions indicated by the symbol $^\ominus$? *(2 marks)*

3. State the equation, including state symbols, that accompanies the enthalpy change of formation of propene, $C_3H_6(g)$. *(2 marks)*

4. Calculate the enthalpy change for the decomposition of magnesium nitrate shown below.
$Mg(NO_3)_2(s) \rightarrow MgO(s) + 2NO_2(g) + \frac{1}{2}O_2(g)$
Given the information shown in the table:

Substance	$\Delta H_f^\ominus\,kJ\,mol^{-1}$
$Mg(NO_3)_2(s)$	−791
$MgO(s)$	−602
$2NO_2(g)$	−33

(2 marks)

4.5 Enthalpy changes of combustion
Specification reference: 3.1.4

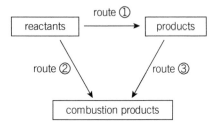

▲ **Figure 1** *Using enthalpy of combustion values: route 1 = route 2 − route 3*

Key term

The standard enthalpy of combustion, ($\Delta_c H^\ominus$): The enthalpy change when one mole of a compound is completely burnt in oxygen under standard conditions.

Summary questions

1. Define the term *standard enthalpy change of combustion*. (1 mark)

2. Write the equation that represents the standard enthalpy change of combustion of propane, $C_3H_8(g)$. (1 mark)

3. Use the information in the table to calculate the standard enthalpy change of formation of ethane. (3 marks)

   $2C(s) + 2H_2(g) \rightarrow C_2H_4(g)$

Substance	$\Delta_c H^\ominus$ kJ mol^{-1}
C(s)	−394
$H_2(g)$	−286
$C_2H_4(g)$	−1409

Using enthalpies of combustion

Hess's law cycles can be used to calculate enthalpy changes that cannot be measured directly. This could happen because the chemicals do not react directly or when a reaction produces several different products. When values for the enthalpy changes of combustion are given in questions the arrow is drawn from the chemicals to the combustion products.

Worked example

Q Calculate the enthalpy change for

$C(s) + 2H_2(g) \rightarrow CH_4(g)$

given that

$\Delta_c H^\ominus C(s) = -394 \text{ kJ mol}^{-1}$

$\Delta_c H^\ominus H_2(g) = -286 \text{ kJ mol}^{-1}$

$\Delta_c H^\ominus CH_4(g) = -890 \text{ kJ mol}^{-1}$

A The unknown enthalpy change $\Delta H^\ominus$

$= -394 \text{ kJ mol}^{-1} + (2 \times -286 \text{ kJ mol}^{-1}) - (-890 \text{ kJ mol}^{-1})$

$= -76 \text{ kJ mol}^{-1}$

▲ **Figure 2** *The Hess's law diagram for the formation of methane.*

Notice how there are two moles of hydrogen in the equation so the value for $\Delta_c H^\ominus$ must be multiplied by two.

Q Calculate the enthalpy change for the hydrogenation of ethene to form ethane

$C_2H_4(g) + H_2(g) \rightarrow C_2H_6(g)$

given that

$\Delta_c H^\ominus C_2H_4(g) = -1409 \text{ kJ mol}^{-1}$

$\Delta_c H^\ominus H_2(g) = -286 \text{ kJ mol}^{-1}$

$\Delta_c H^\ominus C_2H_6(g) = -1560 \text{ kJ mol}^{-1}$

A The unknown enthalpy change $\Delta H^\ominus$

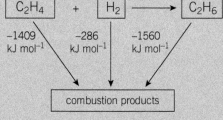

▲ **Figure 3** *The Hess's law diagram for the hydrogenation of ethene.*

$= -1409 \text{ kJ mol}^{-1} - 286 \text{ kJ mol}^{-1} - (-1560 \text{ kJ mol}^{-1})$

$= -135 \text{ kJ mol}^{-1}$

Notice that the method chosen only depends on the values given. Here enthalpy changes of combustion are given so the combustion products are placed at the bottom of the Hess's law diagram and the arrows are drawn downwards.

4.6 Representing thermochemical cycles
Specification reference: 3.1.4

Energetic stability
Enthalpy level diagrams can be used to represent the enthalpy changes of chemical reactions. This enthalpy level diagram (sometimes called an energy level diagram) shows that the products of the reaction, $CO_2(g)$ and $2H_2O(g)$, have a lower energy than the reactants, $CH_4(g)$ and $2O_2(g)$. The products are energetically more stable than the products.

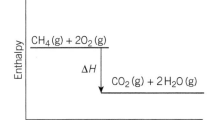

▲ Figure 1 *An exothermic reaction*

Enthalpy values
Enthalpy changes, ΔH, can be worked out from Hess's law cycles. However, to draw enthalpy level diagrams absolute values of the enthalpies of different substances are required.

The enthalpy of elements
The enthalpies of all elements in their standard states (the state that they exist in at 298 K or 25 °C and 100 kPa) are zero.

298 K or 25 °C and 100 kPa are normal temperature and pressure conditions.

Gases such as oxygen normally exist as oxygen molecules, O_2, so the standard state for oxygen is $O_2(g)$.

Thermochemical cycles and enthalpy level diagrams
The enthalpy change of chemical reactions can be shown using thermochemical cycles or as enthalpy level diagrams. Both of these methods are useful ways to tackle problems.

Notice how in thermochemical cycles the arrows show the direction of the change (from the reactants to the products of a chemical reaction) and are drawn diagonally while in enthalpy level diagrams the arrows are drawn vertically.

Revision tip
Some elements exist in several forms in the same state. For example, solid carbon exists as graphite and diamond. These different forms are called allotropes. Under standard conditions the most common form, and therefore the most stable, of the allotropes of carbon is graphite. So graphite is used for the standard state of carbon and is given the symbol C(graphite).

Summary questions

1. State the enthalpy of nitrogen, $N_2(g)$. *(1 mark)*

2. Define what is meant by the term *allotrope*. *(1 mark)*

3. Calculate the standard enthalpy change for the reaction below by using:
   a a thermochemical cycle
   b an enthalpy level diagram
   $2C_2H_6(g) + 7O_2(g) \rightarrow 4CO_2(g) + 6H_2O(g)$
   You are provided with the following data.
   $4C(s) + 6H_2(g) + 7O_2(g) \rightarrow 4CO_2(g) + 6H_2O(g)$
   $\Delta H^\ominus = -3292$ kJ mol^{-1}
   $4C(s) + 6H_2(g) + 7O_2(g) \rightarrow 2C_2H_6(g) + 7O_2(g)$
   $\Delta H^\ominus = -172$ kJ mol^{-1}
   (3 marks)

Worked example
Q Calculate the standard enthalpy change for the reaction below by using:

a a thermochemical cycle

b an enthalpy level diagram

$2NO(g) \rightarrow N_2O_4(g)$

You are provided with the following data.

$N_2(g) + 2O_2(g) \rightarrow 2NO_2(g), \Delta H^\ominus +33$ kJ mol^{-1}

$N_2(g) + 2O_2(g) \rightarrow N_2O_4(g), \Delta H^\ominus +9.0$ kJ mol^{-1}

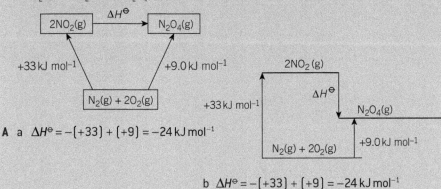

A a $\Delta H^\ominus = -(+33) + (+9) = -24$ kJ mol^{-1}

b $\Delta H^\ominus = -(+33) + (+9) = -24$ kJ mol^{-1}

4.7 Bond enthalpies
Specification reference: 3.1.4

Synoptic link
Bond enthalpies give a measure of the strength of bonds, and can help to predict which bond in a molecule is most likely to break. However, this is not the only factor, the polarity of the bond is also important – see Topic 3.5, Forces acting between molecules, and Topic 13.2, Nucleophilic substitution in halogenoalkanes.

Key term
Bond dissociation enthalpies: The enthalpy change required to break one mole of a covalent bond when all the species are in the gaseous state. This is an endothermic reaction.

Revision tip
Bond enthalpies are always endothermic so they always have a positive sign.

Summary questions

1. Define the term *mean bond enthalpy*. (1 mark)

2. Give the equation that sums up the reaction for the bond dissociation energy of a H–Cl bond. (1 mark)

3. Why does the actual value of bond enthalpy sometimes differ from the mean bond enthalpy for the bond? (1 mark)

4. The enthalpy change for the reaction $CH_4(g) \rightarrow C(g) + 4H(g)$ is $+1664 \text{ kJ mol}^{-1}$. What is the average bond enthalpy for a C–H bond? (1 mark)

Mean bond enthalpy
The mean (average) bond enthalpy is the mean amount of energy required to break one mole of a specified type of covalent bond in a gaseous species.

Why do values calculated from mean bond enthalpies differ from those calculated from Hess's law?
The mean bond enthalpy for a bond is an average value obtained from many molecules. This means that the actual value for the bond in a particular molecule will probably be a little different from the mean value. So values found using calculations involving mean bond enthalpies will sometimes be slightly different from those obtained from other methods such as from using Hess's law.

Using mean bond enthalpies in calculations
Mean bond enthalpy values can be used to work out the enthalpy change of a reaction.

Worked example
Q Calculate the enthalpy change when methane, $CH_4(g)$ is burnt.

Bond	Average bond enthalpy / kJ mol^{-1}
C–H	+413
O=O	+498
C=O	+805
O–H	+464

$CH_4(g) + 2O_2(g) \rightarrow CO_2(g) + 2H_2O(g)$

A Energy required to break the existing bonds

$4 \times \text{C–H} = 4 \times 413 \text{ kJ mol}^{-1} = 1652 \text{ kJ mol}^{-1}$

$2 \times \text{O=O} = 2 \times 498 \text{ kJ mol}^{-1} = 996 \text{ kJ mol}^{-1}$

Total = 2648 kJ mol^{-1}

Energy released when the new bonds are formed

$2 \times \text{C=O} = 2 \times 805 \text{ kJ mol}^{-1} = 1610 \text{ kJ mol}^{-1}$

$4 \times \text{O–H} = 4 \times 464 \text{ kJ mol}^{-1} = 1856 \text{ kJ mol}^{-1}$

Total = 3466 kJ mol^{-1}

The net energy change = $2648 \text{ kJ mol}^{-1} - 3466 \text{ kJ mol}^{-1} = -818 \text{ kJ mol}^{-1}$

Chapter 4 Practice questions

1 Ethene, C_2H_4 reacts with bromine, Br_2 to form 1,2-dibromoethane.

The equation for the reaction is shown below.

$$H_2C=CH_2 (g) + Br-Br(g) \rightarrow Br-CH_2-CH_2-Br(g)$$

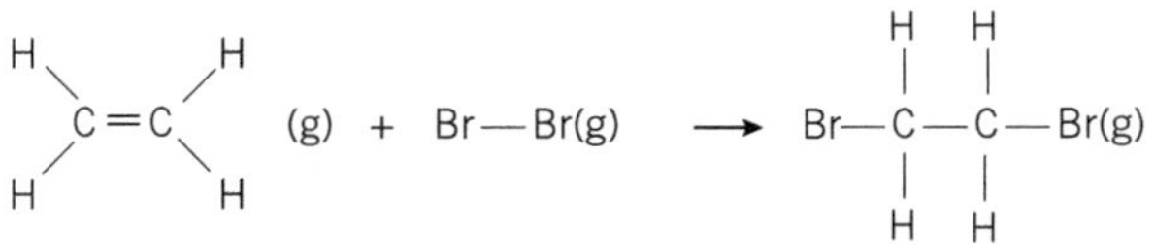

The values for bond enthalpy for a selection of bonds are shown in the margin.

Bond	Mean bond enthalpy / kJ mol^{-1}
C=C	+612
C–H	+413
C–Br	+285
Br–Br	+193
C–C	+348

a Why are bond enthalpy values always endothermic? *(1 mark)*

b Calculate the bond enthalpy change for the reaction between ethene and bromine. *(3 marks)*

c What is the significance of the sign for the bond enthalpy change? *(1 mark)*

2 A student placed 100 cm³ of 1.00 mol dm⁻³ copper(II) sulfate solution into a polystyrene cup and added an excess of magnesium powder.

The equation for the reaction is shown below.

$Mg(s) + CuSO_4(aq) \rightarrow MgSO_4(aq) + Cu(s)$

The temperature of the solution increased from 21.0 °C to 32.0 °C

The specific heat capacity of the solution, $c = 4.18 \, J\,g^{-1}\,K^{-1}$

The density of all the solutions = 1.00 g cm⁻³

a Why did the student add magnesium powder? *(1 mark)*

b Define the term *specific heat capacity*. *(2 marks)*

c Calculate the energy change in the reaction. *(2 marks)*

d Calculate the amount, in mol, of copper(II) sulfate used in the reaction. *(1 mark)*

e Calculate the enthalpy change of the reaction. Give your answer to three significant figures. *(2 marks)*

f What is the major source of error in this experiment? What could the student do to reduce the effect of this problem? *(2 marks)*

3 During combustion 2.00 g of ethanol, C_2H_5OH heated 250 cm³ of water by 30 °C.

The specific heat capacity of the solution, $c = 4.18 \, J\,g^{-1}\,K^{-1}$

The density of water = 1.00 g cm⁻³

a Write an equation to represent the complete combustion of ethanol. Include state symbols. *(2 marks)*

b Calculate the heat energy change for the reaction. *(2 marks)*

c Calculate the amount, in mol, of ethanol burnt in the reaction. Give your answer to three significant figures. *(2 marks)*

d Calculate the enthalpy change of the reaction. Give your answer to three significant figures. *(2 marks)*

4 Consider the equation below which shows the enthalpy change of formation of butane, C_4H_{10}.

$4C(s) + 5H_2(g) \rightarrow C_4H_{10}(g)$

a State Hess's law. *(1 mark)*

b Suggest why it would not be possible to measure the enthalpy change of formation of butane directly. *(1 mark)*

31

5.1 Collision theory
5.2 Maxwell–Boltzmann distribution
Specification reference: 3.1.5

> **Key term**
>
> **Activation energy E_a:** The minimum collision energy that particles must have to react.
>
> - Different reactions have different activation energies.
> - The lower the activation energy the larger the number of particles that can react at any temperature.

Collision theory

The rate of a chemical reaction is a measure of how quickly a reactant is used up or of how quickly a product is made.

You can use the collision theory to understand how the conditions used affect the rate of a chemical reaction.

- For a reaction to occur particles must collide.
- When the particles collide they must have enough energy to break the existing bonds.
- When the particles collide they must collide in the correct orientation so that the reactive parts of the molecules come together. This is particularly important for large molecules. As a result only a small proportion of collisions between particles result in a reaction.

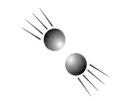

▲ **Figure 1** *Particles must collide before they can react.*

The effect of changing the temperature

- As you increase the temperature of a sample, you increase the kinetic energy of the particles.
- As the temperature increases, the particles move faster so they collide more often.
- Also when the particles do collide more of the particles will have enough energy to react. As a result the higher the temperature the larger the number of particles that can react.

However, the particles will only react if they collide in the correct orientation.

Maxwell–Boltzmann distributions

These diagrams are used to represent the energy of the particles in a sample of a gas at a given temperature.

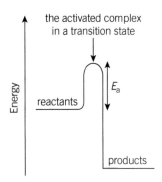

▲ **Figure 2** *The energy change in an exothermic reaction*

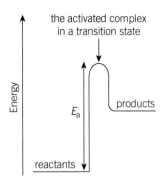

▲ **Figure 3** *The energy change in an endothermic reaction*

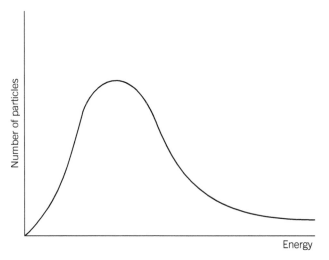

▲ **Figure 4** *The energy distribution curve for a sample of gas at a particular temperature.*

Kinetics

Activation energy and Maxwell–Boltzmann distributions

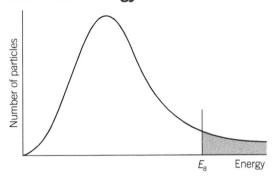

▲ Figure 5 Activation energy

The activation energy is the minimum collision energy that particles must have to react.

- For a reaction to occur the particles must collide and these collisions must have energy equal to or greater than the activation energy. If there is not enough energy when the particles collide then the particles will not react.

- Only the small proportion of particles in the shaded part of the distribution curve have enough energy to react.
- Notice that the activation energy is drawn towards the right of the peak of the distribution curve. If more than half the particles had enough energy to react, the reaction would be too fast to be controlled safely.

Revision tip
- Notice that the distribution curve is not symmetrical.
- Most of the particles have an energy which falls within quite a narrow range, with few particles having much more or much less energy.
- The line does not cross the x-axis at higher energy. In fact it would only do so at infinity.
- The line starts at the origin, showing that none of the particles has no energy.
- The total area under the distribution curve represents the total number of gas particles.

Maxwell–Boltzmann distributions and temperature

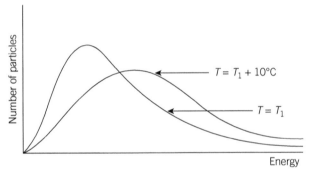

▲ Figure 6 The Maxwell–Boltzmann distributions at different temperatures

Revision tip
Notice how the range of energies that the particles have increases as the temperature increases.

- The average energy of the particles increases.
- The peak of the distribution curve moves to the right as the average energy increases.
- The distribution curve becomes flatter because the total number of particles remains the same.

Summary questions

1. What is the activation energy of a reaction? *(1 mark)*

2. How does increasing the temperature affect the energy of the molecules? *(1 mark)*

3. Describe how the Maxwell–Boltzmann distribution of energies diagram changes if the temperature is decreased by 10 °C. *(2 marks)*

4. Why does increasing the temperature increase the rate of a chemical reaction? *(2 marks)*

5.3 Catalysts

Specification reference: 3.1.5

Catalysts

- A catalyst increases the rate of a chemical reaction but is not used up itself during the reaction.
- Catalysts work by providing an alternative reaction pathway which has lower activation energy. As a result at any given temperature more particles will have energy greater than or equal to the activation energy. As a result the reactions happen faster (the rate of reaction increases).

Catalysts and Maxwell–Boltzmann distributions

Revision tip
Adding a catalyst does not change the distribution curves but it does mean that more particles have enough energy to react.

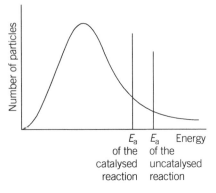

▲ **Figure 1** *A catalysed reaction has a lower activation energy*

Worked example

Q The graph below represents the Maxwell–Boltzmann distribution of energies at a particular temperature.

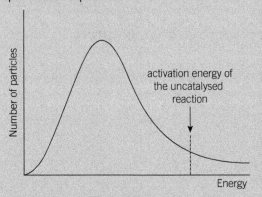

State the effect of adding a catalyst on:

a the energy of the particles

b the activation energy

c the rate of reaction.

A a Adding a catalyst will not affect the energy of the particles.

b Adding a catalyst will provide an alternative reaction pathway with a lower activation energy.

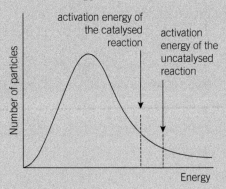

c The rate of reaction will increase because more particles now have enough energy to react.

Summary questions

1 How does adding a catalyst increase the rate and the yield of a chemical reaction? *(2 marks)*

2 Use a Maxwell–Boltzmann distribution of energies diagram to explain how adding a catalyst increases the rate of a chemical reaction. *(2 marks)*

Chapter 5 Practice questions

1 Maxwell–Boltzmann distribution diagrams can be used to show the distribution of energy of the molecules in a sample of a gas.

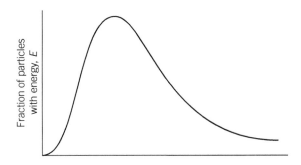

   a Label the *x*-axis of the Maxwell–Boltzmann graph. *(1 mark)*

   b The temperature of the sample of gas is increased by 10 °C. Draw a new line on the graph to show the distribution of energy of the molecules at this higher temperature. *(2 marks)*

   c Explain, in terms of the particles, why increasing the temperature of the sample of a gas can increase the rate of a chemical reaction. *(3 marks)*

2 Consider the Maxwell–Boltzmann distribution diagram below.

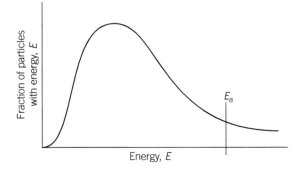

   a Chemical reactions take place when particles collide. Explain, in terms of the particles involved, two reasons why most collisions do not result in a chemical reaction taking place. *(2 marks)*

   b On the diagram above draw a line to show the effect of adding a catalyst on the activation energy of the reaction. *(2 marks)*

   c Catalysts can increase the rate of a chemical reaction. Explain how a catalyst can increase the rate of a chemical reaction. Use your diagram to help you explain your answer. *(2 marks)*

3 a Explain the term *collision theory*. *(2 marks)*

   b Suggest two ways that the collision frequency of the molecules in a gaseous sample could be increased. *(2 marks)*

6.1 The idea of equilibrium
6.2 changing the conditions of an equilibrium reaction
Specification reference: 3.1.6

Reversible reactions and dynamic equilibria

Many reactions are reversible; they can go forwards or backwards. For example:

$$SO_2(g) + \tfrac{1}{2}O_2(g) \rightleftharpoons SO_3(g)$$

- The $\rightleftharpoons$ sign means that the reaction is reversible.
- If the forward reaction is exothermic the enthalpy change will have a negative sign. The reverse reaction will be endothermic and the enthalpy change will have a positive sign.
- If a reversible reaction takes place in a closed system where nothing can enter or leave then an equilibrium can be reached.
- At equilibrium there is a balance between the reactants and the products and the concentrations of the products and reactants stay the same. This does not mean that the concentration of reactants is equal to the concentration of the products.
- The forward and backward reactions have not stopped. It is just that at equilibrium the rate of forward reaction is equal to the rate of backward reaction. As a result the equilibrium is described as being dynamic.

Le Chatelier's principle

Le Chatelier's principle helps us to predict the effect of changing factors that affect the position of equilibrium.

The effect of changing the concentration on the position of equilibrium

If a reaction involving solutions is at equilibrium then changing the concentration will affect the position of equilibrium. The position of equilibrium shifts to minimise the change. If the concentration of one of the reactants is increased it will shift the position of equilibrium towards the right so more product will be made.

The effect of changing temperature on the position of equilibrium

If a reaction is at equilibrium then changing the temperature may affect the position of equilibrium. The position of equilibrium will shift to minimise the change in temperature. As a result if the temperature is increased the position of equilibrium will shift in the endothermic direction.

The effect of changing the pressure on the position of equilibrium

If a reaction that involves gases is at equilibrium then changing the pressure may affect the position of equilibrium. The position of equilibrium shifts to minimise the change. If the pressure is increased it will shift the position of equilibrium towards the side that has fewer gas molecules. If the pressure is decreased it will shift the position towards the side with more gas molecules.

> **Key term**
>
> **Le Chatelier's principle:** Le Chatelier's principle states that when the conditions of a dynamic equilibrium are changed then the position of equilibrium will shift to minimise the change.

> **Revision tip**
>
> Catalysts increase the rate of chemical reaction by offering an alternative reaction pathway with a lower activation energy. Catalysts do not affect the position of equilibrium.

> **Summary questions**
>
> 1. What does the sign $\rightleftharpoons$ mean? *(1 mark)*
>
> 2. What does Le Chatelier's principle state? *(1 mark)*
>
> 3. How does adding a catalyst to a chemical reaction affect the rate of reaction and the position of equilibrium? Explain your answers. *(2 marks)*

6.3 Equilibrium reactions in industry
Specification reference: 3.1.6

Equilibrium conditions
The following conditions apply to all chemical equilibria:
- The reactants and products must be in a closed system (where nothing can enter or leave).
- Equilibrium can be reached from either direction (you can start with reactants or products and the same equilibrium position will be reached provided the same conditions are used).
- The equilibrium is dynamic meaning reactants are reacting to form new products and products are reacting together to form the reactants again but the rate of forwards reaction is equal to the rate of backwards reaction.

Equilibria and chemical processes
The Haber process
Ammonia, NH_3, is produced by the Haber process. The reaction is reversible and the forward reaction is exothermic.

$$N_2(g) + 3H_2(g) \rightleftharpoons 2NH_3(g)$$

- An iron catalyst is used to increase the rate of reaction. This allows a reaction to be carried out at a reasonable temperature.
- The catalyst does not affect the position of equilibrium, so it has no effect on the yield of ammonia at equilibrium.
- The forward reaction between nitrogen and hydrogen is exothermic. Increasing the temperature increases the rate of reaction but decreases the yield of ammonia (increasing the temperature shifts the position of equilibrium in the endothermic direction). As a result a compromise temperature is used. This gives a reasonable rate and a reasonable yield of ammonia.
- Increasing the pressure increases the yield of ammonia (increasing the pressure shifts the position of equilibrium towards the products side, which has fewer gas particles).

The Contact process
Sulfuric acid is produced by the Contact process. The reaction is reversible and the forward reaction is exothermic.

$$SO_2(g) + \tfrac{1}{2}O_2(g) \rightleftharpoons SO_3(g)$$

> **Revision tip**
> The macroscopic (large-scale) properties of the system are constant. For example, the concentration or colour of the equilibrium mixture stay the same.

> **Revision tip**
> Ammonia can be used to produce fertilisers and explosives.

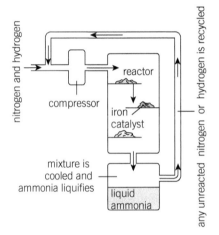

▲ Figure 1 *The Haber process*

Summary questions
1. What is a closed system? *(1 mark)*
2. What is the catalyst used in the Haber process and how does it affect the rate of reaction and the yield of ammonia made? *(3 marks)*

6.4 The equilibrium constant, K_c

Specification reference: 3.1.6

> **Key term**
>
> **Chemical equilibrium:** When the rate of forward reaction is equal to the rate of backward reaction so that the concentration of the reactants and products remain constant.

> **Maths skill**
>
> For the reaction
>
> $H_2(g) + I_2(g) \rightleftharpoons 2HI(g)$
>
> $K_c = \dfrac{(\text{mol dm}^{-3})^2}{(\text{mol dm}^{-3})(\text{mol dm}^{-3})}$
>
> = no units
>
> The units cancel out so there are no units

> **Worked example:**
>
> **Q** For the reaction
>
> $N_2O_4(g) \rightleftharpoons 2NO_2(g)$
>
> Write the expression for the equilibrium constant, K_c and state its units.
>
> **A** $K_c = \dfrac{[NO_2(g)]^2}{[N_2O_4(g)]}$
>
> Units = $\dfrac{(\text{mol dm}^{-3})(\text{mol dm}^{-3})}{(\text{mol dm}^{-3})}$
>
> = mol dm^{-3}

> **Summary questions**
>
> 1. Write the expression for the equilibrium constant, K_c for
>   a. $2NO_2(g) \rightleftharpoons 2NO(g) + O_2(g)$
>   b. $2HCl(g) \rightleftharpoons H_2(g) + Cl_2(g)$
>   c. $2SO_2(g) + O_2(g) \rightleftharpoons 2SO_3(g)$ *(3 marks)*
>
> 2. State the units for the equilibrium constant, K_c for
>   a. $2NO_2(g) \rightarrow 2NO(g) + O_2(g)$
>   b. $2HCl(g) \rightarrow H_2(g) + Cl_2(g)$
>   c. $2SO_2(g) + O_2(g) \rightarrow 2SO_3(g)$ *(3 marks)*

Reversible reactions

Many reactions are reversible and do not go to completion. Eventually, a mixture that contains both reactants and products is formed. The symbol $\rightleftharpoons$ is used to indicate that the reaction is reversible. If a reaction is carried out in a closed system, equilibrium may be reached.

Homogeneous equilibrium

A homogeneous equilibrium is when all the reactants and products in the equilibrium mixture are in the same physical state or phase, for example, all the reactants and products are gases.

The equilibrium constant, K_c

At equilibrium the concentration of the reactants and of products remains constant.

Hydrogen, H_2, and iodine, I_2, can react together to form hydrogen iodide, HI. However, the reaction is reversible and hydrogen iodide molecules can decompose to form hydrogen and iodine. These reactions can be summed up by the equation below:

$H_2(g) + I_2(g) \rightleftharpoons 2HI(g)$

Eventually equilibrium is reached in which the concentration of the two reactants and of the products remains constant.

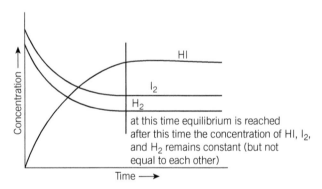

▲ **Figure 1** Reaching equilibrium

At equilibrium the concentrations of hydrogen, iodine, and hydrogen iodide are constant and are linked together by the equilibrium constant, K_c.

For the reaction

$H_2(g) + I_2(g) \rightleftharpoons 2HI(g)$

$K_c = \dfrac{[HI(g)]^2}{[H_2(g)][I_2(g)]}$

The square brackets in the expression indicate that the concentrations are measured in units of mol dm^{-3}

Units for K_c

The units for K_c vary and have to be worked out for each equilibrium system.

6.5 Calculations using equilibrium constant expressions

Specification reference: 3.1.6

Calculating the equilibrium constant 1

If the concentration of the reactants and products present at equilibrium are known the value for the equilibrium constant, K_c, can be calculated quite easily by substituting the values of concentration into the expression for the equilibrium constant.

Calculating the equilibrium constant 2

In some questions the number of moles of reactants and products and the total volume of the equilibrium mixture are given. K_c can still be calculated but first, the concentration of the reactants and products must be deduced using the expression:

Concentration (mol dm^{-3}) = amount (mol) / volume (dm^3). Then the values for the concentration can be substituted into the expression for the equilibrium constant, K_c.

Calculating the equilibrium constant 3

The hardest examples occur when the initial amounts of the reactants are given together with the equilibrium amount of one substance. Although K_c can still be found, several steps are required. First the equilibrium amount of each substance must be deduced. Then the equilibrium concentrations must be calculated. Finally the equilibrium concentrations can be substituted into the expression for the equilibrium constant, K_c.

Worked example:

Q $N_2(g)$, $H_2(g)$ and $NH_3(g)$ exist in equilibrium in a closed system.

$$N_2(g) + 3H_2(g) \rightleftharpoons 2NH_3(g)$$

The concentration of the reactants and products were found to be:

$N_2(g) = 0.0200$ mol dm^{-3}

$H_2(g) = 0.0100$ mol dm^{-3}

$NH_3(g) = 0.0800$ mol dm^{-3}

Calculate the value of the equilibrium constant. Include the units in your answer.

A $K_c = \dfrac{[NH_3(g)]^2}{[N_2(g)][H_2(g)]^3}$

$= \dfrac{(0.0800 \times 0.0800)}{0.0200 \times (0.0100 \times 0.0100 \times 0.0100)}$

$= \dfrac{0.0064}{2.00 \times 10^{-8}}$

$= 320\,000$ mol^{-2} dm^6

Summary questions

1. $H_2(g)$, $Cl_2(g)$ and $HCl(g)$ exist in equilibrium in a closed system. The total volume is 1.00 dm^3.

   $H_2(g) + Cl_2(g) \rightleftharpoons 2HCl(g)$

   The concentration of the reactants and products at equilibrium were found to be:

   $H_2(g) = 0.0500$ mol dm^{-3}
   $Cl_2(g) = 0.100$ mol dm^{-3}
   $HCl(g) = 0.800$ mol dm^{-3}

   Calculate the value of K_c. Give your answer to three significant figures and include the units. *(3 marks)*

2. Consider the equation below.

   $N_2O_4(g) \rightleftharpoons 2NO_2(g)$

   A chemist allowed 0.600 moles of N_2O_4 to decompose in a 500 cm^3 flask.
   The equilibrium mixture was analysed and found to contain 0.250 moles of N_2O_4.
   Calculate the value of K_c. Give your answer to three significant figures and include the units. *(5 marks)*

6.6 The effect of changing conditions on equilibria

Specification reference: 3.1.6

> **Synoptic link**
>
> Look back at Topic 5.1, Collision theory, to revise activation energy for reactions.
>
> The equilibrium constant can also be calculated for gases using partial pressures. This equilibrium constant has the symbol K_p.

Le Chatelier's principle

If a reversible reaction is carried out in a closed system eventually an equilibrium can be reached. However, if the conditions are changed the system is no longer in equilibrium and the position of equilibrium can be shifted.

Changes in pressure

If a reaction that involves gases, and is in dynamic equilibrium, is subject to a change in pressure the position of equilibrium will shift so as to minimise the increase in pressure.

> **Worked example:**
>
> Q Many industrial processes involve reversible reactions.
>
> The Contact process is used in the manufacture of sulfuric acid. In the Contact process SO_2 reacts with O_2 to produce SO_3.
>
> $2SO_2(g) + O_2(g) \rightleftharpoons 2SO_3(g)$
>
> State and explain the effect of increasing the total pressure on the position of equilibrium.
>
> A There are 3 gaseous molecules on the left-hand side of the equation and 2 gaseous molecules on the right-hand side of the equation.
>
> Increasing the total pressure will shift the position of equilibrium to the right-hand side as it is the side with the fewer gaseous molecules and will decrease the total pressure.

Changes in concentration

If a reaction in dynamic equilibrium is subject to a change in concentration of reactants or products the position of equilibrium will shift so as to minimise the change in concentration.

Changes in temperature

If a reaction in dynamic equilibrium is subject to a change in temperature the position of equilibrium will shift so as to minimise the change in temperature.

> **Worked example:**
>
> Q Consider the equation below
>
> $N_2(g) + 3H_2(g) \rightleftharpoons 2NH_3(g) \quad \Delta H = -92 \text{ kJ mol}^{-1}$
>
> State and explain the effect of decreasing the temperature on the position of equilibrium.
>
> A The forward reaction is exothermic, whilst the backward reaction is endothermic.
>
> Decreasing the temperature will shift the position of equilibrium to the right-hand side as this is the exothermic direction.

The equilibrium constant, K_c

If a reversible reaction is exothermic in the forward direction, increasing the temperature will decrease the equilibrium constant, K_c. If a reversible reaction is endothermic in the forward direction increasing the temperature will increase the equilibrium constant, K_c.

> **Summary questions**
>
> 1 Consider the equation below
> $N_2(g) + 3H_2(g) \rightleftharpoons 2NH_3(g)$
>   State and explain the effect of a decrease in total pressure on the position of equilibrium.
>   *(2 marks)*
>
> 2 Consider the equation below
> $X(g) + 3Y(g) \rightleftharpoons 2Z(g)$,
> $\Delta H = -100 \text{ kJ mol}^{-1}$
>   State and explain the effect of increasing the temperature on the position of equilibrium.
>   *(2 marks)*
>
> 3 Consider the equation below
> $E(g) + 2F(g) \rightleftharpoons G(g)$
> $\Delta H = -120 \text{ kJ mol}^{-1}$
>   State and explain the effect of increasing the temperature on the value of the equilibrium constant, K_c.
>   *(2 marks)*

Chapter 6 Practice questions

1 Nitrogen and hydrogen react together to form ammonia.

$N_2(g) + 3H_2(g) \rightleftharpoons 2NH_3(g)$, $\Delta H = -92 \text{ kJ mol}^{-1}$

A chemist mixes 3.00 moles of nitrogen with 6.00 moles of hydrogen in a 1.00 dm³ flask. At equilibrium there was found to be 2.00 moles of NH_3.

  a Deduce the equilibrium concentrations of nitrogen and hydrogen. *(2 marks)*

  b Write the expression for the equilibrium constant, K_c, for this equilibrium. *(1 mark)*

  c Calculate the value of the equilibrium constant. Give your answer to an appropriate number of significant figures and include the units. *(3 marks)*

  d A chemist increases the temperature of the reaction. State and explain what happens to the value for the equilibrium constant as the temperature is increased. *(1 mark)*

  e A chemist increases the pressure of the reaction mixture. State and explain what happens to the position of equilibrium as the pressure is increased. *(1 mark)*

2 The reaction between sulfur dioxide and oxygen to produce sulfur trioxide is reversible. The equation for the reaction is shown below.

$2SO_2(g) + O_2(g) \rightleftharpoons 2SO_3(g)$

The forward reaction is exothermic.

The system was allowed to reach equilibrium in a closed system.

  a Define the term *closed system*. *(1 mark)*

  b The equilibrium can be described as being dynamic. What can be said about the rate of forward reaction in a dynamic equilibrium? *(1 mark)*

  c State and explain the effect of increasing the temperature on the position of equilibrium. *(2 marks)*

  d Vanadium (V) oxide can be used to catalyse the reaction between sulfur dioxide and oxygen. State and explain the effect of adding a vanadium (V) oxide catalyst on the position of equilibrium. *(2 marks)*

3 Ammonia can be produced from nitrogen and hydrogen using the Haber process. The equation for the reaction is shown below.

$N_2(g) + H_2(g) \rightleftharpoons 2NH_3(g)$

The forward reaction is exothermic.

The reaction is carried out at a temperature of 450 °C and a pressure of 200 atmospheres. A finely divided iron catalyst was used.

  a State and explain the effect of increasing the temperature on: *(2 marks)*

    i the rate of reaction

    ii the position of equilibrium.

  b State and explain the effect of increasing the pressure on: *(2 marks)*

    i the rate of reaction

    ii the position of equilibrum.

  c State and explain the effect of using the iron catalyst on: *(2 marks)*

    i The rate of reaction

    ii The position of equilibrium.

7.1 Oxidation and reduction
7.2 Oxidation states
Specification reference: 3.1.7

> **Key term**
>
> **Oxidation:** The loss of electrons.

> **Key term**
>
> **Reduction:** The gain of electrons.

Oxygen and hydrogen

The term *redox* is short for reduction–oxidation and these two reactions always occur together. Historically the term *oxidation* was used for reactions in which oxygen is added to a substance and the term reduction was used for reactions in which oxygen was removed from a substance.

Consider the equation below:

$PbO + H_2 \rightarrow Pb + H_2O$

Oxygen is added to the hydrogen, H_2, to form water, H_2O, so the hydrogen is oxidised. At the same time oxygen is removed from the lead(II) oxide, PbO to form lead, Pb so the lead(II) oxide is reduced. In a similar way historically the term *reduction* was used for reactions in which hydrogen was added to a substance.

$CH_3CHO + 2[H] \rightarrow CH_3CH_2OH$

Hydrogen, H, is added to the ethanal, CH_3CHO, to form ethanol, CH_3CH_2OH, so the ethanal is reduced. If the reaction were reversed and hydrogen was removed from ethanol to form ethanal the ethanol would be oxidised.

Oxidation and reduction

As one species loses electrons another species must gain these electrons. As a result oxidation and reduction must always occur together. These reactions are often called 'redox reactions'.

Oxidising and reducing agents

- An oxidising agent is a species that oxidises another substance by removing electrons from it.
- Oxidising agents are themselves reduced during the reaction.
- Common oxidising agents include oxygen, chlorine, potassium dichromate(VI), and potassium manganate(VII).
- A reducing agent is a species that reduces another substance by adding electrons to it.
- Reducing agents are themselves oxidised during the reaction.
- Common reducing agents include Group 1 and 2 metals, hydrogen, carbon, and carbon monoxide.

Assigning oxidation states

- Any pure element has an oxidation state of zero.
- Any monatomic ion has an oxidation state equal to the charge on the ion.
- In compounds, Group 1 metal atoms have an oxidation state of +1.
- In compounds, Group 2 metal atoms have an oxidation state of +2.
- In compounds, the most electronegative element fluorine always has an oxidation state of −1.

Oxidation, reduction, and redox reactions

- In compounds, hydrogen has an oxidation state of +1 unless it is in a metal hydride when it has an oxidation state of −1.
  - The total oxidation state of an uncharged molecule is 0.
  - The total oxidation state of any polyatomc ion is equal to the charge of the ion.

Examples

The oxidation state of hydrogen in H_2O is +1.

The oxidation state of hydrogen in KH is −1.

- In compounds, the oxidation state of oxygen is −2 unless it is with fluorine when it has an oxidation state of +2 or is part of peroxide when it has an oxidation state of −1.

Examples

The oxidation state of oxygen in H_2O is −2.

The oxidation state of oxygen in F_2O is +2.

The oxidation state of oxygen in H_2O_2 is −1.

- If a molecule is neutral overall, for example CO_2, then the sum of the oxidation states of the atoms in the molecule must be zero.
- Where a molecular ion has an overall charge, for example CO_3^{2-}, then the sum of the oxidation states of the atoms in the molecule must equal the overall charge of the ion.

Summary questions

1. What is the oxidation state of oxygen in O_2? *(1 mark)*

2. What is the oxidation state of sulfur in SO_4^{2-}? *(1 mark)*

3. State the oxidation number of iodine in:
   a KIO_3 b KI   c KIO_4 *(3 marks)*

Worked example

Q State the oxidation number of sulfur in:

a H_2SO_4
b H_2S
c SO_2

A

a KIO_3 So I must be +5
   $+1 \times 1$ -2×3
   $= +1$ $= -6$

b KI   So I must be −1
   $+1 \times 1$
   $= +1$

c KIO_4 So I must be +7
   $+1 \times 1$ -2×4
   $= +1$ $= -8$

Maths skill

Make sure you state the oxidation state of each atom. For example in $K_2Cr_2O_6$ the oxidation state of Cr is +6.

7.3 Redox equations

Specification reference: 3.1.7

> **Key term**
>
> **Oxidation:** The loss of electrons.

> **Key term**
>
> **Reduction:** The gain of electrons.

> **Revision tip**
>
> **OIL RIG**
>
> The acronym OIL RIG is a useful way of remembering that
>
> Oxidation
>
> Is the
>
> Loss of electrons
>
> Reduction
>
> Is the
>
> Gain of electrons

Half-equations

The reaction for the oxidation of magnesium can be summed up by the equation

$2Mg + O_2 \rightarrow 2MgO$

Half-equations can be used to show what happens to each of the reactants in a redox reaction.

$2Mg \rightarrow 2Mg^{2+}$

$O_2 + 4e^- \rightarrow 2O^{2-}$

Using oxidation states to identify oxidation and reduction reactions

When magnesium is added to copper sulfate solution a displacement reaction takes place.

The overall equation for the reaction is

$$Mg + CuSO_4 \rightarrow MgSO_4 + Cu$$

- The oxidation state of uncombined magnesium is zero; in $MgSO_4$ it is +2. The oxidation state has gone up so the magnesium is oxidised.
- The oxidation state of copper is +2 in $CuSO_4$ and zero in uncombined copper. The oxidation state has gone down so the copper has been reduced.
- We can show the electron transfer by splitting the overall equation into two half-equations.

$$Mg \rightarrow Mg^{2+} + 2e^- \text{ (oxidation)}$$

$$Cu^{2+} + 2e^- \rightarrow Cu \text{ (reduction)}$$

The magnesium is oxidised and the copper is reduced. The magnesium has reduced the copper so it is a reducing agent. The copper sulfate has oxidised the magnesium so it is an oxidising agent. Take care to identify the name of the reagent rather than just the copper as the oxidising agent. The sulfate ions are not involved in the reaction and are referred to as being spectator ions.

> **Summary questions**
>
> 1. Define the terms *oxidation* and *reduction*. *(2 marks)*
>
> 2. Consider the two half-equations below.
> $Fe^{2+} + 2e^- \rightarrow Fe$
> $Zn \rightarrow Zn^{2+} + 2e^-$
>   Combine these two half-equations to give the overall equation for the reaction. *(1 mark)*
>
> 3. Consider the equation below:
> $Zn + 2HCl \rightarrow ZnCl_2 + H_2$
>   Split the overall equation into two half-equations. *(2 marks)*
>
> 4. Magnesium reacts with chlorine to form magnesium chloride.
>   a Write the overall equation for the reaction.
>   b Give the two half-equations for the reaction. *(3 marks)*

Chapter 7 Practice questions

1 Iron is a transition element and forms ions with several different oxidation states.
   a Deduce the oxidation state of the iron in each of these substances. *(2 marks)*
      i $FeCl_3$
      ii $FeCO_3$
   b Give the full name of each of these compounds. *(2 marks)*
      i $FeCl_3$
      ii $FeCO_3$

2 Complete the table below to show the oxidation state of sulfur in each of the substances.

Substance	Oxidation state of S
H_2S	
SO_2	
SO_3	
H_2SO_4	
S_8	

(5 marks)

3 Consider the reaction below
   $Fe(s) + CuSO_4(aq) \rightarrow Cu(s) + FeSO_4(aq)$
   a What is the oxidation state of the reagent iron? *(1 mark)*
   b Use oxidation states to explain which species are oxidised and reduced in this reaction. *(2 marks)*
   c Name the oxidising agent in this reaction. *(1 mark)*

4 Magnesium reacts with hydrochloric acid in a redox reaction.
   The equation for the reaction is shown below.
   $Mg(s) + 2HCl(aq) \rightarrow MgCl_2(aq) + H_2(g)$
   a Describe what you would see during this reaction. *(2 marks)*
   b Explain, in terms of electron transfer, which species is oxidised and which is reduced in this reaction. *(2 marks)*

5 Complete the table below to show the oxidation state of the element required in each substance.

	Oxidation state
Mg in $MgCO_3$	
Cu in Cu_2O	
Cl in $NaClO_4$	

(3 marks)

6 Which of these statements describes an oxidising agent?
   A A substance that is oxidised.
   B An element that is oxidised.
   C A substance that gains electrons during a chemical reaction
   D A substance that does not change oxidation state during a chemical reaction. *(1 mark)*

7 Identify the oxidation state of Cl in the compound $NaClO_3$
   A −1   C −5
   B +5   D +7   *(1 mark)*

8.1 The Periodic Table
8.2 Trends in the properties of elements of Period 3

Specification reference: 3.2.1

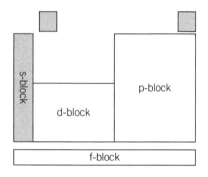

▲ **Figure 1** *Blocks of the Periodic Table*

> **Synoptic link**
> To revisit covalent bonding, look back at Topic 3.2, Covalent bonding.

> **Synoptic link**
> To revisit metallic bonding, look back at Topic 3.3, Metallic bonding.

> **Synoptic link**
> To revisit intermolecular bonding, look back at Topic 3.5, Forces acting between molecules.

> **Key term**
> **Periodic:** Recurs regularly.

s-block, p-block, and d-block elements

Elements that belong to the s-block have their highest energy electrons in s-orbitals.

For example, lithium, Li $1s^2\ 2s^1$, is an s-block element.

p-block elements have their highest energy electrons in p-orbitals

For example, nitrogen, N $1s^2\ 2s^2\ 2p^3$ is a p-block element.

d-block elements

d-block elements have their highest energy electrons in d-orbitals.

For example, titanium, Ti $1s^2\ 2s^2\ 2p^6\ 3s^2\ 3p^6\ 4s^2\ 3d^2$ is a d-block element.

Groups

A group is a vertical column in the Periodic Table. Elements in the same group have the same number of electrons in their outer shell and so have similar chemical properties.

Periods

A period is a horizontal row in the Periodic Table.

Periodicity

Examining the Periodic Table reveals many patterns in the properties of the elements.

Across each period the metals are found on the left and the non-metals on the right. This is a regularly recurring pattern and is an example of periodicity.

Melting points and boiling points

Trends in melting point and boiling point across Period 3 are related to the change in bonding and structure of the elements.

▼ **Table 1** *Table of the melting points and boiling points of Period 3 elements*

Element	Melting point/K	Boiling point/K
Na	371	1156
Mg	922	1380
Al	933	2740
Si	1683	2628
P	317	553
S	392	718
Cl	172	238
Ar	84	87

Periodicity

Sodium, magnesium, and aluminum are metals and have a giant metallic structure. Metallic bonds are strong so their melting points and boiling points are quite high.

Silicon is a semi-metal, which forms a giant covalent (macromolecular) structure in which each silicon atom is bonded to four other silicon atoms by very strong covalent bonds so the melting point and boiling points of silicon are very high.

Phosphorus, P_4, sulfur, S_8, and chlorine, Cl_2, are non-metals that form simple molecules. The van der Waals' bonds between these molecules are weak so these elements melt and boil at low temperatures.

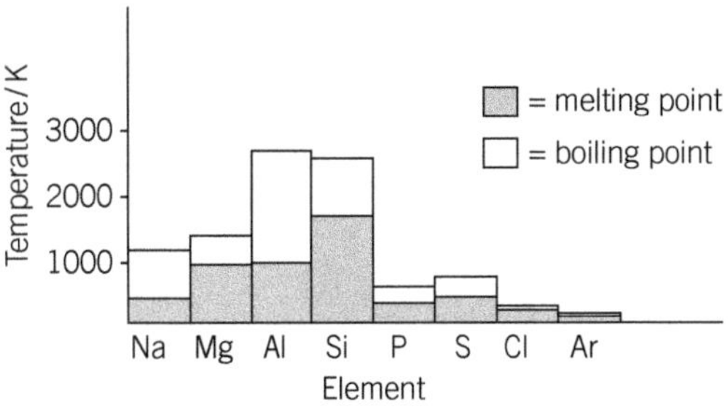

▲ **Figure 2** *Melting points and boiling points of the Period 3 elements*

▼ **Table 2** *Period 3 elements*

Element	Type of bonding structure	Force broken on melting
sodium	giant metallic	metallic bond
magnesium	giant metallic	metallic bond
aluminium	giant metallic	metallic bond
silicon	giant covalent	covalent bond
phosphorus	simple covalent, P_4 units	van der Waals' force
sulphur	simple covalent, S_8 units	van der Waals' force
chlorine	simple covalent, Cl_2 units	van der Waals' force
argon	monatomic	van der Waals' force

More about structures of some Period 3 elements

Sodium, magnesium, and aluminum are all metals. In magnesium each atom has two electrons in its outer shell. Both these electrons are delocalised leaving a magnesium ion with a 2+ charge. The metallic bonds in magnesium are stronger than the metallic bonds in sodium (with only one delocalised electron and ions with a 1+ charge) so the melting point of magnesium is higher than the melting point of sodium.

Revision tip
Try practising drawing a small section of silicon using shaded and dotted lines to show its three dimensional structure.

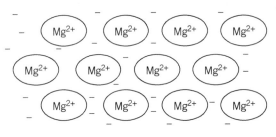

▲ **Figure 3** *The metallic bonding in magnesium*

The bonding in silicon is giant covalent with a network of covalent bonds throughout the structure. Melting or boiling silicon requires the breaking of covalent bonds so requires a lot of energy.

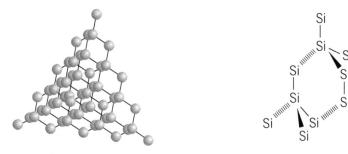

▲ **Figure 4** *Silicon has a giant covalent structure*

Phosphorus, sulfur, and chlorine are all simple molecules with van der Waals' forces between them. This force strength increases with the increasing size of the molecule (number of electrons) so the melting or boiling point increases in the order argon, chlorine, phosphorus, and sulfur.

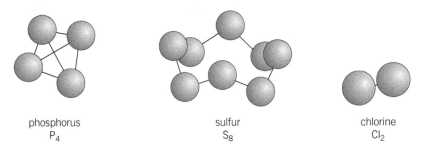

phosphorus P_4 sulfur S_8 chlorine Cl_2

▲ **Figure 5** *Phosphorus, sulfur, and chlorine have a simple molecular structure*

Summary questions

1. Give the full electron configuration for each of the elements below and state the block of the Periodic Table that each element belongs to.
   a boron   b fluorine
   c neon   d potassium
   e iron *(5 marks)*

2. Define the term *periodic*. *(1 mark)*

3. What is the bonding and structure in a chlorine b sodium? *(2 marks)*

4. Sulfur and silicon both contain covalent bonds. Explain why sulfur melts at 392 K but silicon melts at 1683 K. *(2 marks)*

47

8.3 More trends in the properties of elements of Period 3
8.4 A closer look at ionisation energies
Specification reference: 3.2.1

> **Revision tip**
> You should be able to write the electron configuration for all the atoms in Period 3. Practice reading this information from the Periodic Table. Remember that elements with outer electrons in the s energy level can be classified as s-block elements and those with outer electrons in the p energy level can be classified as p-block elements.

> **Key term**
>
> **First ionisation energy:** First ionisation energy is the energy required to remove one electron from each atom in one mole of gaseous atoms forming one mole of ions with a single positive charge.

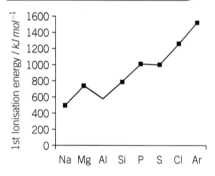

▲ **Figure 1** *First ionisation energy for Period 3 elements*

> **Summary questions**
>
> 1 Define the term *first ionisation energy*. *(2 marks)*
>
> 2 State and explain the trend in the atomic radii of Period 3 elements. *(2 marks)*
>
> 3 Give the equation that represents the first ionisation energy of magnesium. *(2 marks)*
>
> 4 Why does sulfur have a lower first ionisation energy than phosphorus? *(2 marks)*

Atomic radius

The **atomic radius** of the elements decreases across Period 3 as the number of protons in the nucleus increases across the period but the corresponding extra electrons are placed into the same shell. This means that the electrostatic attraction of the positive nucleus on the negative electrons increases slightly across the period, making the atomic radius slightly smaller.

Electron configurations and first ionisation energy

It is important that you can describe and explain the graph showing the first ionisation energy for the elements in Period 3. There is a general increase across Period 3 as the number of protons in the nucleus steadily increases, leading to a greater charge on the nucleus. As a result there is a greater attraction between the nucleus and the outer electron so more energy is required to remove the electron from the gaseous atom.

Why are the increases in ionisation energy not regular?

There is a slight decrease between magnesium and aluminium because the outer electron in aluminium is in a p sub-level which is higher in energy than the outer electron in magnesium which is in an s sub-level. So aluminium loses its outer electron slightly more easily than aluminium loses its outer electron.

▼ **Table 1** *The electron configurations of magnesium and aluminium*

Element	Electron configuration	Spin diagrams				
		1s	2s	2p	3s	3p
Mg	$1s^2\,2s^2\,2p^6\,3s^2$	↑↓	↑↓	↑↓ ↑↓ ↑↓	↑↓	
Al	$1s^2\,2s^2\,2p^6\,3s^2\,3p^1$	↑↓	↑↓	↑↓ ↑↓ ↑↓	↑↓	↑

There is another slight decrease between phosphorus and sulfur. The outer electron configuration of sulfur has a pair of electrons in one of the p-orbitals. There is repulsion between the electrons so less energy is needed to remove an electron from this pair so sulfur has a lower first ionisation energy than phosphorus.

▼ **Table 2** *The electron configurations of phosphorus and sulfur*

Element	Electron configuration	Spin diagrams				
		1s	2s	2p	3s	3p
P	$1s^2\,2s^2\,2p^6\,3s^2\,3p^3$	↑↓	↑↓	↑↓ ↑↓ ↑↓	↑↓	↑ ↑ ↑
S	$1s^2\,2s^2\,2p^6\,3s^2\,3p^4$	↑↓	↑↓	↑↓ ↑↓ ↑↓	↑↓	↑↓ ↑ ↑

Chapter 8 Practice questions

1. Complete the table below to show the bonding and structure of the Period 3 elements.

Element	Bonding	Structure
Na		
Mg	metallic	
Al		giant metallic lattice
Si	covalent	
P_4		simple molecules
S_8	covalent	
Cl_2		
Ar	van der Waals'	Monatomic

(2 marks)

2. Sodium, aluminium, and chlorine belong to Period 3 of the Periodic Table.
   a. In terms of their structure and bonding, explain why chlorine has a lower melting point than sodium. (3 marks)
   b. In terms of their structure and bonding explain why aluminium has a higher melting point than sodium. (2 marks)

3. Phosphorus, P_4, sulfur, S_8, and chlorine, Cl_2, exist as simple molecules. Predict which of these elements has the highest melting point. Explain your answer. (3 marks)

4. Identify the Period 3 element that has the following successive ionisation energies. Explain your answer. (1 mark)

	1st	2nd	3rd	4th	5th	6th
Ionisation energy / kJ mol^{-1}	577	1820	2740	11600	14800	18400

5. In terms of structure and bonding, explain why silicon has a very high melting point. (3 marks)

6. Which Period 3 element has the second largest atomic radius?
   A Na
   B Ar
   C Cl
   D Mg (1 mark)

 Go further

1. Give the equation that represents the first ionisation energy of nitrogen. (2 marks)
2. Explain why oxygen has a lower first ionisation energy than nitrogen. (3 marks)

9.1 The physical and chemical properties of Group 2

Specification reference: 3.2.2

Element	Atomic radius / nm
Mg	0.145
Ca	0.194
Sr	0.219
Ba	0.253

Element	First ionisation energy / kJ mol^{-1}
Mg	738
Ca	590
Sr	550
Ba	503

Revision tip
Notice how in Figure 1 the calcium ions have a 2+ charge.

Introducing the alkaline earth metals

Group 2 metals have two electrons in their outer shell. They react to form ions that have a 2+ charge.

$$Mg \rightarrow Mg^{2+} + 2e^-$$

Trend in atomic radius

Down the group atoms have an extra shell of electrons and become larger. As a result the atomic radii of the elements increase down the group.

Trend in ionisation energy

- Down the group the outer electrons are lost more easily.
- Although the number of protons increases down the group, the atomic radii also increase so the distance between the nucleus and the outer electrons increases.
- There is also more shielding by the inner electrons. As a result the value for the first ionisation energy decreases down the group.

Trend in melting points

Group 2 elements have reasonably high melting points due to metallic bonding.

Element	melting point / °C
Mg	649
Ca	839
Sr	769
Ba	729

Metallic bonding

In metals the electrons in the outer shell of atoms are delocalised. This leads to positive metal ions and negatively charged delocalised electrons.

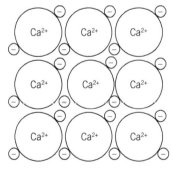

▲ **Figure 1** *The metallic bonding in calcium metal*

- Metallic bonding is the electrostatic attraction between the positive ions and the negative delocalised electrons.
- Down the group the size of the metal ions increases so the strength of the metallic bonding decreases.
- Less energy is required to overcome the forces of attraction. As a result melting points generally decrease down the group.

Group 2, the alkaline earth metals

Trend in reactivity
- The reactivity increases down the group.
- Group 2 elements are strong reducing agents because they can lose electrons quite easily.
- Down the group the elements become stronger reducing agents.

How do Group 2 elements react with water?

Group 2 metal atoms are good reducing agents. Their reaction with water is an example of a redox reaction.

- The Group 2 metal atom is oxidised.

$$M \rightarrow M^{2+} + 2e^-$$

- The hydrogen in water is reduced.

Magnesium reacts slowly with water to form magnesium hydroxide and hydrogen.

$$Mg(s) + 2H_2O(l) \rightarrow Mg(OH)_2(aq) + H_2(g)$$

Magnesium hydroxide is sparingly soluble.

Magnesium reacts quickly with steam.

$$Mg(s) + H_2O(g) \rightarrow MgO(s) + H_2(g)$$

Calcium, strontium, and barium all react vigorously with water to form a metal hydroxide and hydrogen.

Example

$$Ca(s) + 2H_2O(l) \rightarrow Ca(OH)_2(aq) + H_2(g)$$

A solution of calcium hydroxide looks cloudy as it is only slightly soluble in water and the undissolved calcium hydroxide forms a white suspension.

Trend in the solubilities of the Group 2 hydroxides

The solubility of Group 2 hydroxides increases down the group.

Group 2 hydroxide	Solubility (g per 100 cm³ of water)	Solubility
$Mg(OH)_2$	0.0012	↓
$Ca(OH)_2$	0.113	
$Sr(OH)_2$	0.410	
$Ba(OH)_2$	2.57	

Trend in the solubilities of the Group 2 sulfates

The solubility of Group 2 sulfates decreases down the group.

Group 2 sulfate	Solubility (g per 100 cm³ of water)	Solubility
$MgSO_4$	25.5	↑
$CaSO_4$	0.24	
$SrSO_4$	0.013	
$BaSO_4$	0.00022	

Key term

Solubility: The mass of the substance that will dissolve in 100 cm³ of water.

Summary questions

1. State the trend in the solubilities of the Group 2 sulfates. *(1 mark)*

2. State and explain the trend in the atomic radii of the Group 2 elements. *(1 mark)*

3. State and explain the general trend in the melting points of the Group 2 elements. *(1 mark)*

Chapter 9 Practice questions

1. State and explain the change in the first ionisation energy of Group 2 elements. *(4 marks)*

2. A student added an excess of magnesium powder to a solution of hydrochloric acid.

   The equation for the reaction is shown below:

   $Mg(g) + 2HCl(aq) \rightarrow MgCl_2(aq) + H_2(g)$

   State and explain which species are oxidised and reduced in this reaction. *(2 marks)*

3. Choose one answer in each section. *(3 marks)*

   a Down Group 2 the size of the atoms:

   increases/decreases/stays the same.

   b Down Group 2 the melting point of the elements:

   increases/decreases/stays the same.

   c Down Group 2 the strength of the metallic bonding in the elements:

   increases/decreases/stays the same.

4. Magnesium hydroxide is sometimes called milk of magnesia. Give a medical use of magnesium hydroxide and explain why it works. *(2 marks)*

5. Give the equation for the second ionisation energy of strontium. *(2 marks)*

6. Give the full electron configuration of: *(3 marks)*

   a Mg

   b Ca

   c Ca^{2+}

7. Which of these Group 2 metals has the second smallest atoms? *(1 mark)*

   A Be

   B Mg

   C Sr

   D Ba

8. Which equation represents the first ionisation energy of magnesium? *(1 mark)*

   A $Mg \rightarrow Mg^+ + e^-$

   B $Mg(g) \rightarrow Mg(g)^+ + e^-$

   C $Mg^+ + e^- \rightarrow Mg$

   D $Mg(g) \rightarrow Mg(g)^{2+} + 2e^-$

10.1 The halogens
Specification reference: 3.2.3

Introducing the halogens

The elements in Group 7 of the Periodic Table are often called the halogens.

Halogen atoms have seven electrons in their outer shell.

When halogen atoms react they can gain an electron to form a halide ion which has a −1 charge.

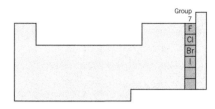

▲ **Figure 1** *Group 7 elements are typical non-metals*

Trend in atomic radius

Down the group atoms have an extra shell of electrons and become larger. As a result the atomic radii (size) of the elements increase down the group.

Trends in electronegativity

Element	Electronegativity
F	4.0
Cl	3.0
Br	2.8
I	2.5

Down the group the atoms have a higher atomic number so the greater number of protons should be able to attract the electrons more. However, down the group:

- The atoms have more shells of electrons. This means there is more shielding and less attraction between the nucleus and the electrons.
- The atomic radius increases. This results in less attraction between the nucleus and the electrons in the covalent bond. As a result down the group the halogens become less electronegative.

Trends in boiling point

At room temperature:

- Fluorine is a yellow gas.
- Chlorine is a pale green gas.
- Bromine is a brown liquid.
- Iodine is a dark grey solid.

Notice how the boiling points of the halogens increase down the group.

- The halogens exist as diatomic molecules, for example, chlorine exists as chlorine molecules, Cl_2.
- This means that there are strong covalent bonds within the halogen molecules but only very much weaker van der Waals' forces of attraction between halogen molecules.
- The strength of the van der Waals' forces depends on the number of electrons in the molecules.
- Down the group the number of electrons in the halogen molecules increases, so the strength of the Van der Waals' forces of attraction between molecules increases. As a result down the group the boiling point increases.

> **Synoptic link**
>
> Look back at Topic 3.2, Covalent bonding.

> **Synoptic link**
>
> Look back at Topic 3.5, Forces acting between molecules.

> **Key term**
>
> **Electronegativity:** A way of measuring the ability of an atom to draw the electrons in a covalent bond to itself.

> **Summary questions**
>
> 1 Define the term *electronegativity*. *(1 mark)*
>
> 2 State and explain the trend in atomic radius down Group 7. *(2 marks)*
>
> 3 State and explain the trend in electronegativity down Group 7. *(2 marks)*

10.2 The chemical reactions of halogens
Specification reference: 3.2.3

Key terms

Oxidising agent: A species that oxidises another substance by removing electrons from it.

Oxidation: The loss of electrons.

Reduction: Reduction is the gain of electrons.

Half-equation: A half-equation is either of the two equations that describe each half of a redox reaction.

Trend in the oxidising ability of the halogens

The oxidising ability of a substance is a measure of the strength of an atom to attract and gain an electron.

- The halogens are good oxidising agents.
- Oxidising agents are themselves reduced during the reaction.

When halogen atoms react they gain electrons.

Example

$$Cl_2 + 2e^- \rightarrow 2Cl^-$$

The oxidising ability of the halogens decreases down the group. Down the group the electron that is gained is placed into a shell that is further from the nucleus.

- Down the group the atomic radii increase.
- The amount of shielding increases. As a result the attraction between the nucleus and the electron decreases. As a result the oxidising ability of the halogens decreases down the group. Fluorine is a very powerful oxidising agent.

Displacement reactions

The displacement reactions between halogens and aqueous halides demonstrate the decrease in the oxidising ability of the halogens down the group.

Chlorine is a more powerful oxidising agent than bromine so chlorine oxidises bromide ions.

Example

$$Cl_2(aq) + 2Br^-(aq) \rightarrow 2Cl^-(aq) + Br_2(aq)$$

The half-equations for this reaction are

$$Cl_2(aq) + 2e^- \rightarrow 2Cl^-(aq) \text{ (reduction)}$$

Chlorine is reduced to chloride.

$$2Br^-(aq) \rightarrow + Br_2(aq) + 2e^- \text{(oxidation)}$$

Bromide is oxidised to bromine.

Once the halogen has been displaced it forms a solution. Different halogens form different coloured solutions. These colours can be used to show a reaction has taken place. However, it is difficult to tell bromine and iodine apart when they are dissolved in water. If an organic solvent, such as cyclohexane, is added and the mixture shaken the halogen dissolves in the organic solvent to give a much clearer colour.

▼ **Table 1** *Identifying halogens*

Halogen	Water	Cyclohexane
chlorine, Cl_2	pale green	pale green
bromine, Br_2	orange	orange
iodine, I_2	brown	violet

Summary questions

1. Define the term *oxidising agent*. (1 mark)

2. State and explain the trend in the oxidising ability of the halogens. (3 marks)

3. A chemist bubbles chlorine, Cl_2 gas through a solution of potassium iodide, KI.
   i. Give the equation for the reaction.
   ii. State the colour that would be observed when cyclohexane is added shaken with the reaction mixture.
   iii. State which species has been oxidised and explain your answer. (4 marks)

10.3 Reactions of halide ions
10.4 Uses of chlorine
Specification reference: 3.2.3

Trends in the reducing ability of halide ions

Reducing agents are oxidised during the reaction. The larger the ion the more easily it loses an electron. Down the group the halide ions become increasingly good reducing agents. As a result iodide ions are the strongest reducing agent. Iodide ions are easiest to oxidise.

Identifying halide ions

You can identify halide ions using acidified silver nitrate solution.

- Fluoride ions form silver fluoride, which is a soluble salt so no precipitate is seen.
- The other halide ions form insoluble salts. The colour of the silver salt can be used to identify the halide.

▼ **Table 1** *Identifying halide ions*

Halide ion	Salt formed	Colour of the precipitate
chloride	AgCl	white
bromide	AgBr	cream
iodide	AgI	yellow

The solubility of the silver halides in ammonia can be used to confirm the identity of the halide ions.

▼ **Table 2** *Confirming the identity of halide ions*

Silver halide	Solubility in ammonia
AgCl	Dissolves in dilute ammonia solution.
AgBr	Does not dissolve in dilute ammonia solution but does dissolve in concentrated ammonia solution.
AgI	Does not dissolve even in concentrated ammonia solution.

Chlorine and water treatments

Chlorine reacts with water to form two acids: chloric(I) acid and hydrogen chloride.

$$Cl_2(aq) + H_2O(l) \rightleftharpoons HClO(aq) + HCl(aq)$$

The reaction of chlorine with sodium hydroxide

Chlorine reacts with cold, dilute sodium hydroxide to form sodium chloride, sodium chlorate(I) NaOCl, and water. This is another example of disproportionation. The chlorine is simultaneously oxidised and reduced. Sodium chlorate(I) is used to make household bleaches.

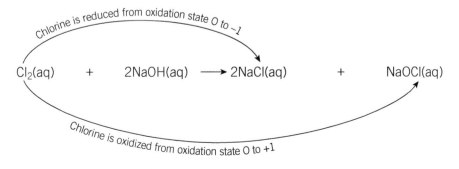

▲ Figure 2

Key term

Reducing agent: A reducing agent is a species which reduces another substance by adding electrons to it.

▲ Figure 1

Revision tip

The oxidation state of chlorine in Cl_2 is 0; in HClO is +1 and in HCl is −1.

This is a disproportionation reaction.

The chlorine is simultaneously oxidised (from 0 to +1) and reduced (from 0 to −1).

Summary questions

1 Give the oxidation number of chlorine in;
 a KCl
 b $NaClO_3$ *(2 marks)*

2 How could you identify bromide ions? *(2 marks)*

3 State and explain the trend in the reducing ability of the halide ions. *(2 marks)*

Chapter 10 Practice questions

1. A student has a sample of a Group 1 halide. The student wants to identify the halide in the sample. Outline what the student could do to identify the halide ion present in the sample. Include any observations that the student should make. *(5 marks)*

2. Choose one answer in each section: *(3 marks)*
   a. Down Group 7 the size of the atoms:
      increases/decreases/stays the same.
   b. Down Group 7 the melting point of the elements:
      increases/decreases/stays the same.
   c. Down Group 2 the strength of the van der Waals' bonding between molecules:
      increases/decreases/stays the same.

3. A halide ion can be displaced by a more reactive halogen. The equation for the reaction between a solution of chlorine and a solution of sodium bromide is shown below.

   $Cl_2(aq) + 2NaBr(aq) \rightarrow Br_2(aq) + 2NaCl(aq)$

   a. What you would observe during this reaction? *(1 mark)*
   b. State which species is oxidised and which species is reduced during the reaction. Explain your answer. *(2 marks)*
   c. Name the oxidising agent in this reaction. *(1 mark)*

4. Which of these halogens is the least electronegative? *(1 mark)*
   A F
   B Cl
   C Br
   D I

5. Which of the options below correctly identifies the oxidation number of the chlorine in each substance.

	KCl	NaOCl	NaClO$_4$
A	+1	+1	+7
B	−1	−1	+3
C	−1	+1	+7
D	−1	+1	−7

   (1 mark)

6. Which of these species is the most powerful reducing agent?
   A F$^-$
   B Cl$^-$
   C Br$^-$
   D I$^-$

   (1 mark)

11.1 Carbon compounds
Specification reference: 3.3.1

The chemistry of carbon

Organic chemistry is the study of compounds containing carbon combined with other elements. Carbon can form rings and chains that can be branched or unbranched. Carbon atoms have four electrons in their outer shell and can from four covalent bonds to other atoms. In addition carbon bonds are quite strong. The ability to form these bonds means that a very large number of organic compounds exist.

Empirical formula

The empirical formula of a compound can be easily deduced from its molecular formula.

The empirical formula of a compound can also be calculated from percentage composition by mass data.

> **Worked example**
>
> Q An organic compound was analysed and found to contain:
>
> 52.2% carbon
>
> 34.8% oxygen
>
> 13.0% hydrogen
>
> A Calculate the empirical formula of the organic compound.
>
> ▼ **Table 1** Empirical formula = C_2OH_6
>
> | Element | C | O | H |
> |---|---|---|---|
> | % by mass | 52.2 | 34.8 | 13.0 |
> | Divide by relative atomic mass | $\frac{52.2}{12.0} = 4.35$ | $\frac{34.8}{16.0} = 2.175$ | $\frac{13.0}{1.0} = 13$ |
> | Divide by smallest value | $\frac{4.35}{2.175} = 2$ | $\frac{2.175}{2.175} = 1$ | $\frac{13}{2.175} = 6$ |

Key terms

Empirical formula: The simplest whole number ratio of the atoms in a compound.

Molecular formula: The actual number of each type of atom in a compound.

Skeletal formula: These are simplified formula where straight lines are drawn to represent carbon–carbon bonds. Hydrogen atoms joined to the carbon atoms are removed but other functional groups are shown.

Displayed formula: A formula showing all the bonds in a compound.

Structural formula: A formula showing in minimum detail the arrangement of the atoms in the molecule.

Molecular formula

The molecular formula of a compound can be easily worked out from its displayed formula.

> **Worked example: What is the molecular formula of this organic compound?**
>
> The compound contains three carbon atoms, six hydrogen atoms and one oxygen atom. The molecular formula of the compound is C_3H_6O.
>
>

Maths skill

Ethene has the molecular formula C_2H_4.

For every carbon atom there are two hydrogen atoms.

Its empirical formula is CH_2.

The molecular formula of a compound is a whole number multiple of its empirical formula. The molecular formula can be calculated using the empirical formula of the compound and the relative molecular mass of the compound.

Introduction to organic chemistry

Displayed formula

The displayed formula is a very useful way of showing the structure of organic compounds as every atom and bond is shown.

A single line – represents a single covalent bond.

A double line = represents a double covalent bond. For example:

The organic compound butane has the molecular formula C_4H_{10}.

Its displayed formula is

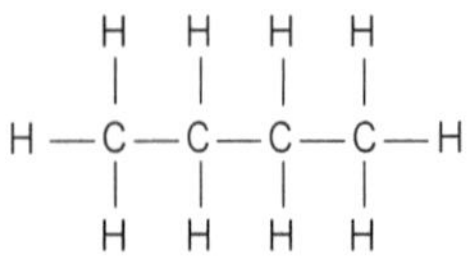

```
    H  H  H  H
    |  |  |  |
H — C— C— C— C — H
    |  |  |  |
    H  H  H  H
```

Skeletal formula

Organic compounds can have very complicated structures and it can become impractical to draw out displayed formula for such complex structures.

Skeletal formulae are a very useful method of representing the molecules in a simplified but precise way. The carbon skeleton of the compound is shown together with any functional groups. For example:

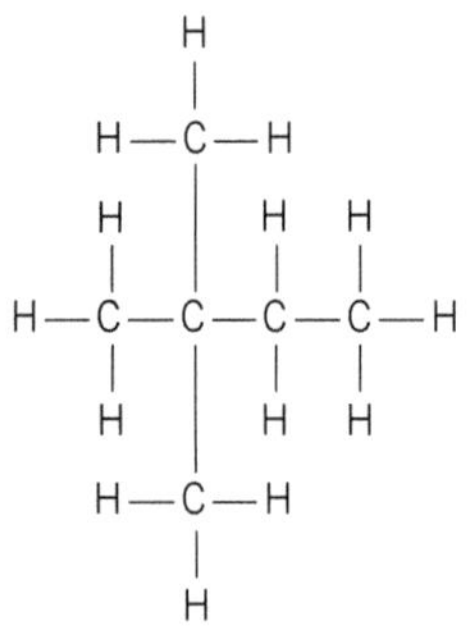

displayed formula 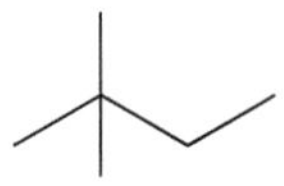   skeletal formula

Structural formula

Molecular formulae can be useful but sometimes two compounds have the same molecular formula. A structural formula is a more useful to chemists. It reveals the structure of the molecule in a very concise way. For example:

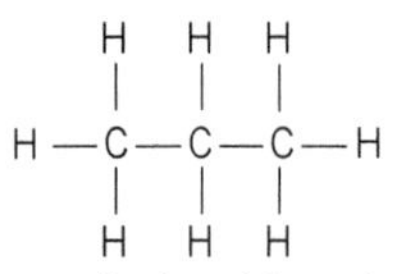

```
    H  H  H
    |  |  |
H — C— C— C — H              CH₃ CH₂ CH₃
    |  |  |
    H  H  H
displayed formula            structural formula
```

Summary question

1. Consider the molecule drawn below.

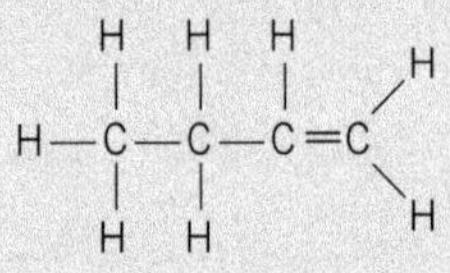

```
    H  H  H
    |  |  |      H
H — C— C— C = C
    |  |         \
    H  H          H
```

a What type of formula is shown in the diagram?
b Give the molecular formula of the compound.
c Give the structural formula of the compound.
d Give the empirical formula of the compound.
e Give the skeletal formula of the compound. *(5 marks)*

11.2 Nomenclature – naming organic compounds

Specification reference: 3.1.1

Naming organic molecules

▼ **Table 1** *Functional groups*

Functional group	Name
alkane	-ane
alkene	-ene
haloalkane	bromo-, chloro-, iodo-
alcohol	-ol

The alkanes

The alkanes are a homologous series which means they have the same functional group. Alkanes have the general formula C_nH_{2n+2}. For straight-chain alkanes the beginning of the name tells you the number of carbon atoms in the chain.

- For example methane has one carbon atom in the chain; ethane has two carbon atoms, and so on.

If the chain is branched then it is said to have a side chain. The side chain is named methyl, ethyl, propyl, etc. depending on the number of carbon atoms it contains. This is then put at the front of the name of the main chain.

- For example methylbutane has a main chain of four carbon atoms and a side chain of one carbon atom.

If there is more than one position that the side chain could attach to the main chain then it is given a number. This is counted from the end of the chain in order to make the number of the side chain as small as possible.

methylbutane
$CH_3CH(CH_3)CH_2CH_3$
C_5H_{12}

2,3-dimethylpentane
$CH_3CH(CH_3)CH(CH_3)CH_2CH_3$
C_7H_{16}

The alkenes

The alkene family is another homologous series. They have a double covalent bond between two of the carbon atoms.

Alkenes have the general formula C_nH_{2n}.

The alkenes are named in the same way as the alkanes except that their name ends in -ene. The position of the C=C double bond is numbered if there is more than one place where it can go in a molecule. The numbering is such that it has the smallest number possible.

Key terms

Homologous series: A group of compounds with the same general formula.

Nomenclature: The system of naming compounds.

Functional group: The part of the molecule that makes it react the way it does.

▼ **Table 2** *Number of carbon atoms*

Number of C atoms	Prefix
1	meth
2	eth
3	prop
4	but
5	pent
6	hex

▼ **Table 3** *Side chains*

number of C atoms	structure	Name
1	-CH$_3$	methyl
2	-CH$_2$CH$_3$	ethyl
3	-CH$_2$CH$_2$CH$_3$	propyl

▼ **Table 4** *Number of identical side chains*

Number of identical side chains	Prefix
2	di
3	tri
4	tetra

Summary question

1 Name the following compounds:
   a C_6H_{14} (unbranched)
   b C_5H_{12} (unbranched)
   c $CH_3CHCHCH_3$
   d $CH_3CH_2CH(CH_2CH_3)CH_2CH_2CH_3$
   e $CH_3CHClCH_2CH_3$
   f $CH_3CH_2CH_2CHClCH_3$
                        (6 marks)

11.3 Isomerism

Specification reference: 3.3.1

> **Synoptic link**
>
> E–Z isomerism is discussed further in Topic 14.1, Alkenes.

> **Key terms**
>
> **Isomerism:** Isomerism occurs when two or more organic molecules have the same molecular formula but different arrangements of atoms.
>
> **Structural isomerism:** Structural isomerism occurs when two or more molecules have the same molecular formula but different structural formulae.

Isomerism

There are two main types of isomerism: structural isomerism and stereoisomerism.

Structural isomerism

There are three types of structural isomerism: positional, functional, and chain.

Positional isomerism

Positional isomerism occurs when a functional group can be in more than one position on the carbon chain.

You must consider positional isomerism when the molecule contains a chain of four or more carbon atoms or when the functional group is a halogen that can be in different positions on the chain.

- In butene (C_4H_8) there are two possible positions for the C=C, between atoms 1 and 2 or between atoms 2 and 3.

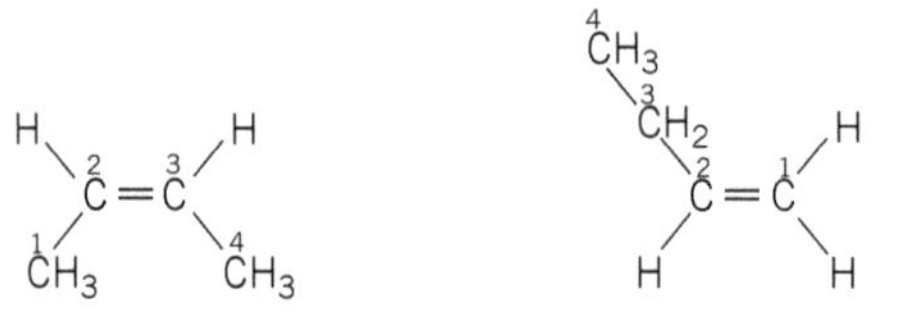

▲ **Figure 1** Positional isomers of C_4H_8

Functional isomerism

Functional isomerism occurs when there are two different functional groups.

propan-1-ol is an alcohol   ethoxymethane is an ether

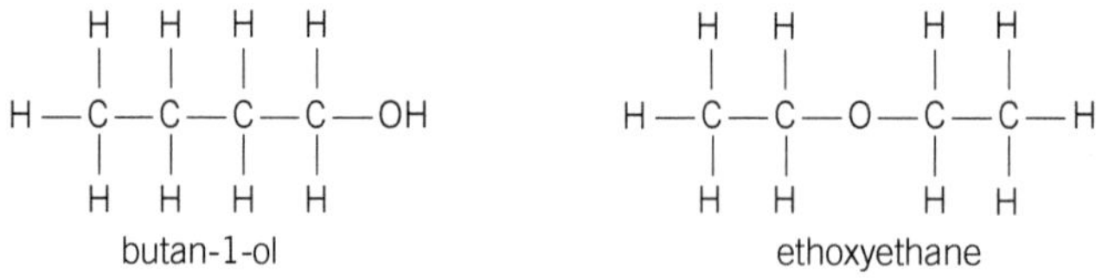

▲ **Figure 2** Butan-1-ol and ethoxyethane are functional isomers

Chain isomerism

Chain isomerism occurs when two or more molecules have the same molecular formula but different arrangements of the carbon chain. Typically this will involve straight chain and branched chain structures.

For example, there are two chain isomers of C_4H_{10} as the carbon atoms can be arranged in a straight chain forming butane or as a chain of three carbon atoms with one side chain forming methylpropane.

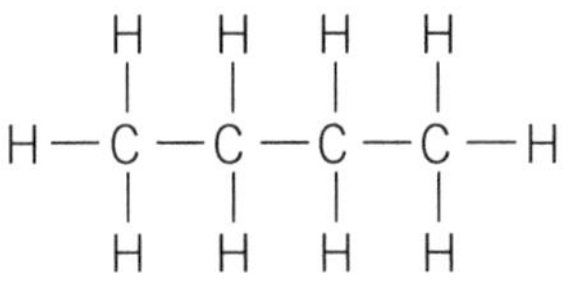

butane

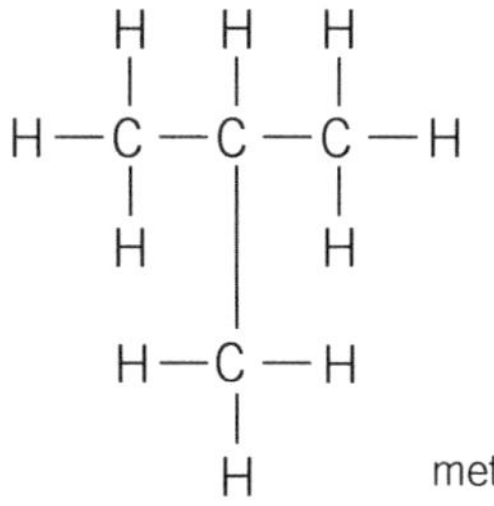

methylpropane

▲ **Figure 3** *Butane and methylpropane are chain isomers*

For C_5H_{12} there are three possible chain isomers; the carbons atoms can be arranged in a straight chain forming pentane, as a chain of four atoms with a side chain forming methylbutane or as a chain of three atoms with two side chains forming dimethylpropane (the prefix di is used to show that there are two methyl groups attached to the central carbon atom).

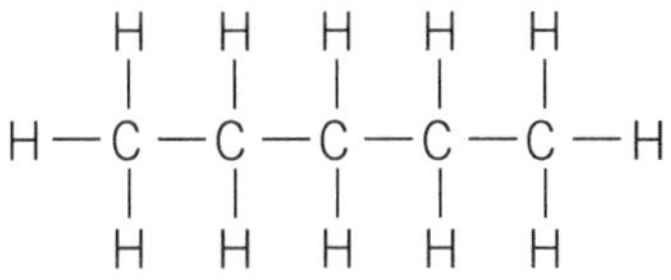

pentane

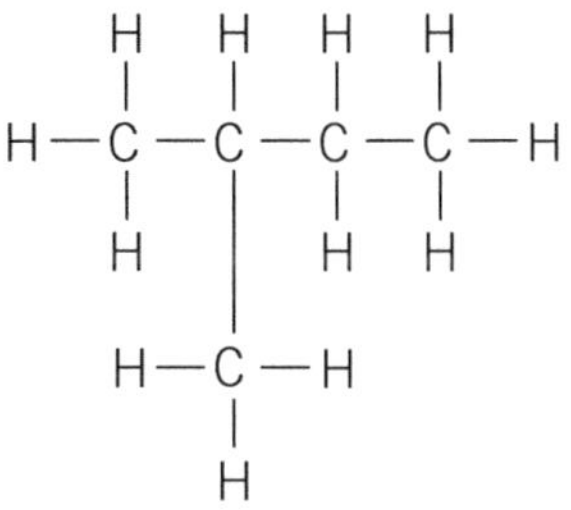

methylbutane

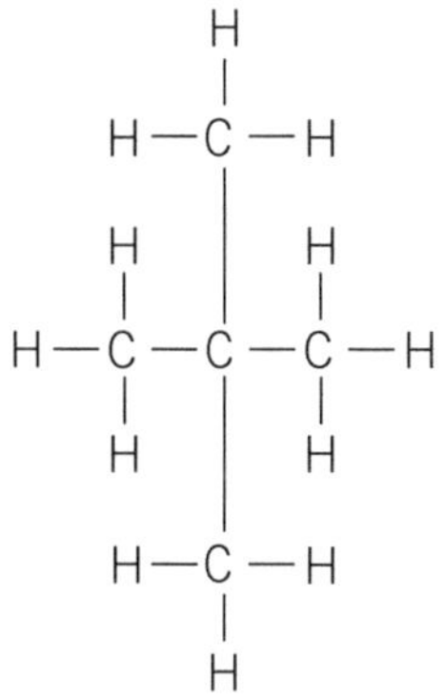

dimethylpropane

▲ **Figure 4** *Chain isomers*

More complex examples

For C_5H_{10} there are five possible isomers. Note that these are a combination of positional and chain isomers.

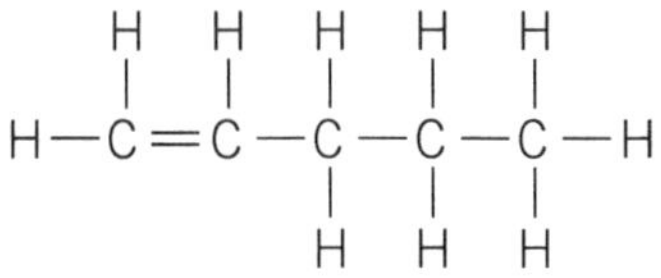

pent-1-ene

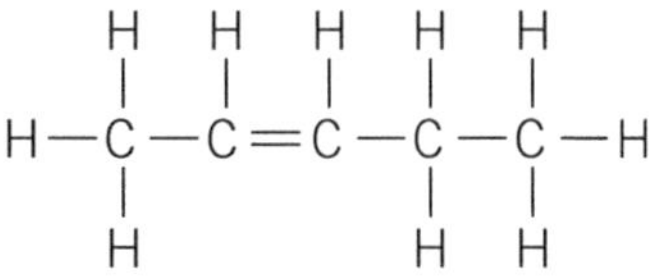

pent-2-ene

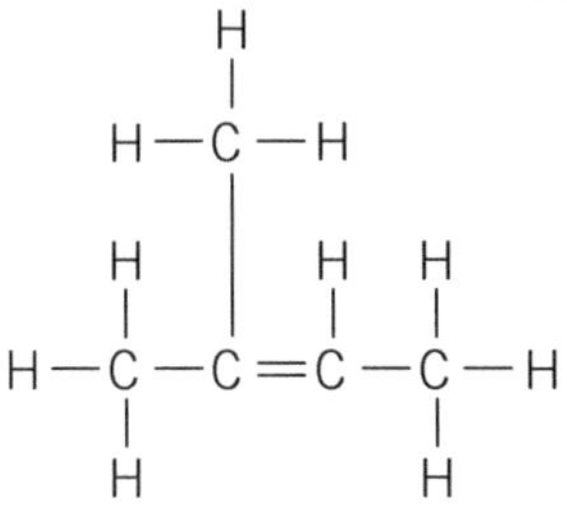

2-methylbut-2-ene

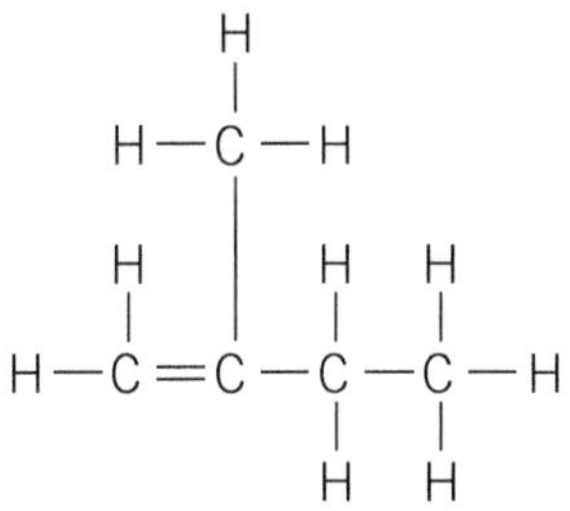

2-methylbut-1-ene

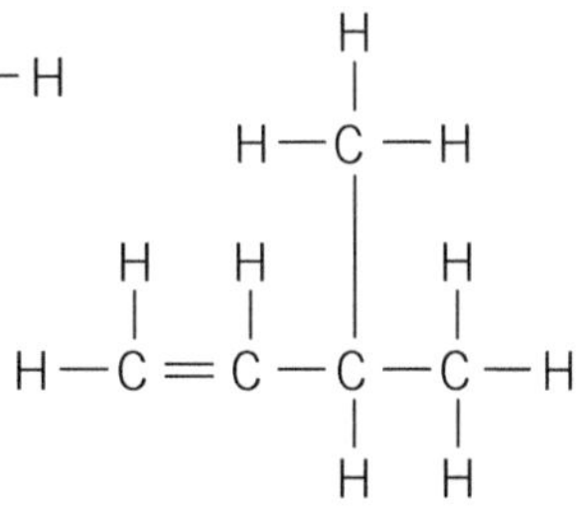

3-methylbut-1-ene

▲ **Figure 5** *Isomers of C_5H_{10}*

Revision tip

It is important that you approach drawing isomers in a systematic way to ensure that you do not miss any out. Name each one when you have drawn it to make sure that you do not draw the same molecule twice.

Summary questions

1. What are the three types of structural isomerism? *(1 mark)*
2. Draw displayed formulae and name all the chain isomers of C_4H_{10}. *(2 marks)*
3. A straight chain organic compound has the molecular formula C_4H_9I. Draw displayed formulae and name the positional isomers of C_4H_9I. *(2 marks)*

Chapter 11 Practice questions

1. An unbranched hydrocarbon **A** has the molecular formula C_4H_8.
   a. Define the term *hydrocarbon*. (*1 mark*)
   b. Draw the displayed formula and name the three possible structure of hydrocarbon **A**. (*3 marks*)

2. Give the skeletal formula and names of the two organic molecules with the molecular formula C_3H_7Br. (*2 marks*)

3. Which of these options shows the functional group in each of the following organic families? (*1 mark*)

	Alkene	Aldehyde	Carboxylic acid
a	C=C	—C(=O)H	—C(=O)O—H
b	C—C	—C(=O)H	—C(=O)OH
c	C=C	—C(=O)H	—C(=O)OH
d	C—C	—C(=O)OH	—C(=O)—

4. The alkanes are a homologous series of hydrocarbons. The table below shows the name and molecular formula of the first five members of the alkane homologous series.

Name	Molecular formula
methane	CH_4
ethane	C_2H_6
propane	C_3H_8
butane	C_4H_{10}
pentane	C_5H_{12}

   a. Define the term *homologous series*. (*2 marks*)
   b. What is the general formula of the alkane homologous series? (*1 mark*)
   c. Give the molecular formula of the tenth member of the alkane homologous series. (*1 mark*)

5. A sample of a hydrocarbon was analysed and found to contain 83.7% carbon. The hydrocarbon has a relative molecular mass of 86.0.
   a. Define the term *empirical formula*. (*1 mark*)
   b. Calculate the empirical formula of the hydrocarbon. (*3 marks*)
   c. Deduce the molecular formula of the hydrocarbon. (*1 mark*)

6. Which of these compounds is a member of the alkane homologous series?
   A C_3H_6 **B** C_2H_2
   C C_3H_8 **D** C_2H_4

7. a. Define the term *structural isomer*. (*1 mark*)
   b. Name the positional isomer of propan-1-ol. (*1 mark*)

8. What is the relative molecular mass of butane?
   A 26   **B** 44
   C 56   **D** 58   (*1 mark*)

12.1 Alkanes
Specification reference: 3.3.2

The alkanes
The alkanes are saturated hydrocarbons. This means that alkanes contain single covalent bonds (not double covalent bonds) and they only contain carbon and hydrogen atoms.

For example:

- The alkane with only 1 carbon atom has $(2 \times 1 + 2) = 4$ hydrogen atoms and the molecular formula CH_4.
- The alkane with 26 carbon atoms has $(2 \times 26 + 2) = 54$ hydrogen atoms and the molecular formula $C_{26}H_{54}$.

> **Revision tip**
> Alkanes have the general formula C_nH_{2n+2}

Naming alkanes
The alkanes all have names ending in -ane. The first part of their name is deduced from the number of carbon atoms in the longest carbon chain.

> **Synoptic link**
> See Topic 11.2, Nomenclature – naming organic compounds, for more details on naming alkanes.

Examples of straight chain alkanes

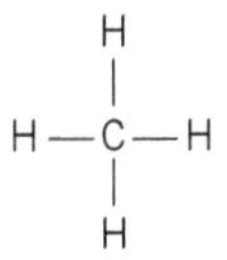

methane
CH_4

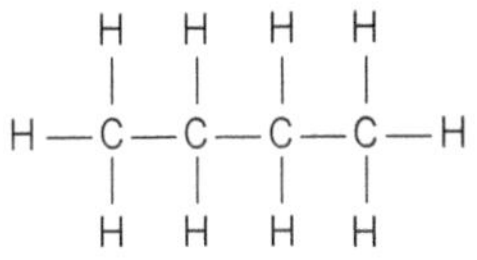

butane
C_4H_{10}

> **Revision tip**
> Notice how cyclic alkanes do not have the same general formula as straight chain alkanes.

Examples of branched alkanes

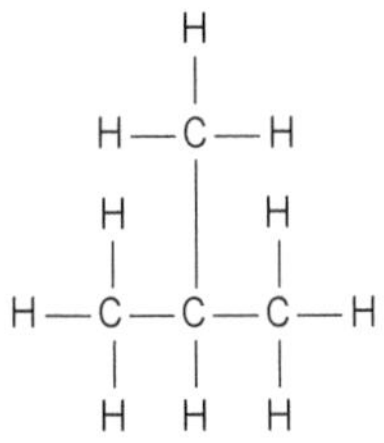

methyl propane
C_4H_{10}

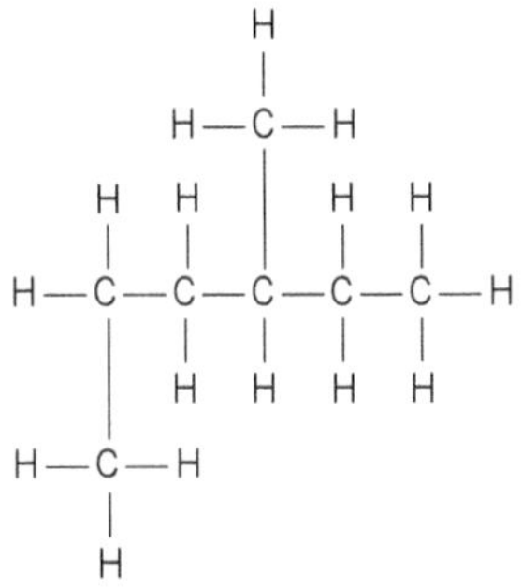

3-methyl-hexane

Examples of ring alkanes
In cyclic or ring alkanes the carbon atoms join together to form a ring of atoms. Ring alkanes have a molecular formula of C_nH_{2n}.

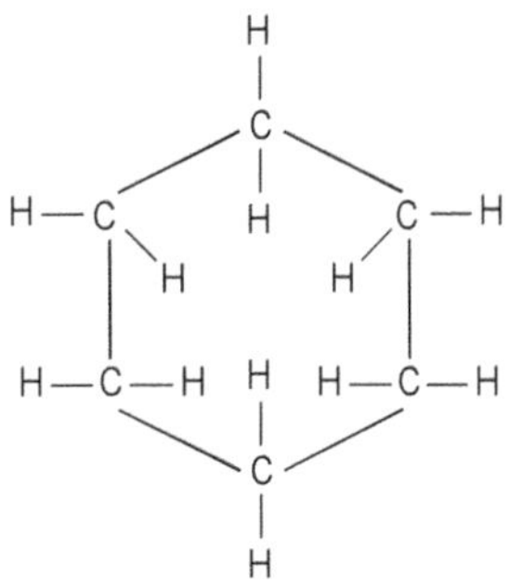

cyclohexane
C_6H_{12}

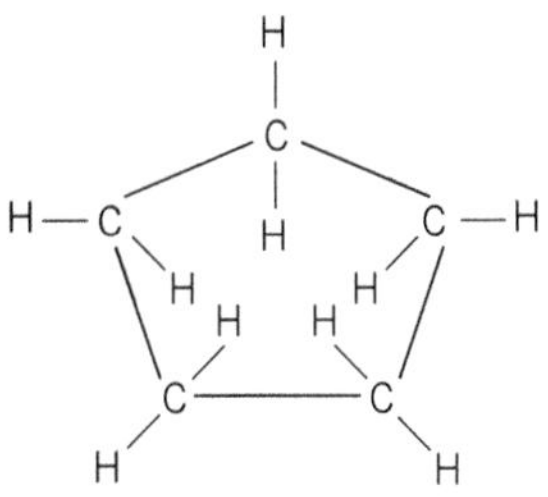

cyclopentane
C_5H_{10}

The properties of alkanes
Polarity
There is very little difference in electronegativity between carbon and hydrogen atoms so alkanes are almost non-polar molecules.

Alkanes

Revision tip
When naming organic molecules look for the longest carbon chain – it does not matter if it is not drawn in a straight line.

Boiling points

The boiling points of alkanes increase as the number of carbon atoms in the molecules increases. Short chain alkanes (1 to 4 carbon atoms) are gases at room temperature. Medium chain alkanes (5 to 17 carbon atoms) are liquid at room temperature. Long chain alkanes (18 or more carbon atoms) are solid at room temperature. As the molecules get larger there are more electrons and therefore stronger van der Waals' forces between the molecules so their melting points and boiling points increase.

Solubilty

Alkanes are almost non-polar molecules so they do not dissolve in polar solvents such as water. However, they will mix with non-polar solvents.

Summary questions

1 What is a saturated hydrocarbon? *(2 marks)*

2 What is the general formula of an alkane? *(1 mark)*

3 Name the straight chain hydrocarbons that have the molecular formula:
   a C_2H_6
   b C_3H_8
   c C_5H_{12} *(3 marks)*

4 Name these organic molecules: *(2 marks)*

   a (six-carbon straight chain alkane structural formula)

   b (branched alkane structural formula)

5 State and explain the trend in the boiling points of the alkane family. *(2 marks)*

6 Why doesn't propane dissolve in water? *(1 mark)*

12.2 Fractional distillation of crude oil
Specification reference: 3.3.2

The origin of crude oil
Crude oil is a very important raw material. It is a fossil fuel that formed millions of years ago from the fossilised remains of dead plants and animals that were buried at high temperatures and pressures below the Earth's surface. Crude oil is a mixture of many substances but mainly consists of branched and unbranched alkane molecules. Small amounts of other substances, such as sulfur, can also be found in crude oil. When crude oils that contain sulfur are burnt sulfur dioxide can be formed. This gas can form acid rain.

Fractional distillation
Crude oil is a mixture and it must be separated into different parts or fractions. Crude oil is a very important source of organic chemicals and this is done on an industrial scale by fractional distillation.

In this process, alkanes are separated into groups with similar boiling points. These groups are called fractions.

The fractionating tower
- The petroleum is heated to around 350 °C so that it forms a vapour.
- The vapour is then passed into the fractionating tower.
- Note that the temperature of the tower falls as it is ascended (this is called a negative temperature gradient).
- The largest hydrocarbons have the highest boiling points so remain as liquids at the bottom of the tower.
- These are tapped off as residue.
- The smallest hydrocarbons have the lowest boiling points so rise up through the tower and leave the top as gases.
- The remaining hydrocarbons rise up the tower until they reach the region that corresponds to their boiling point.
- They then condense at the bubble caps and are removed from the tower as liquids.

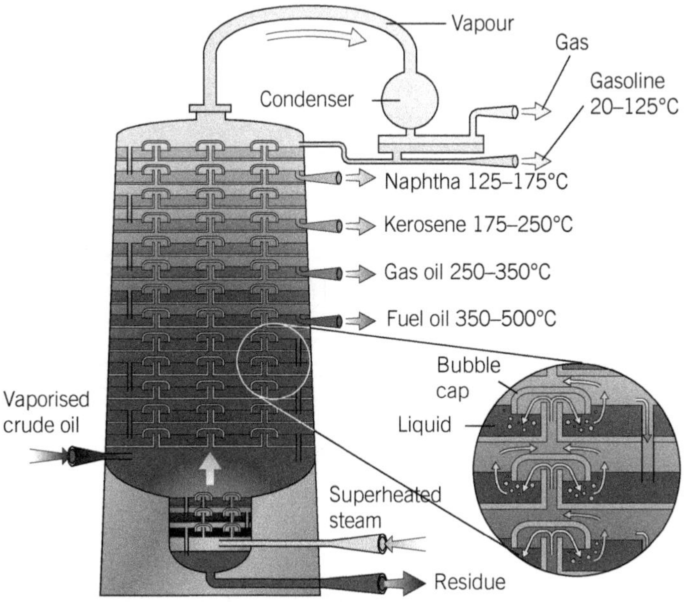

> **Key term**
>
> **Fractional distillation:** The process of separating the crude oil is called fractional distillation. This relies on the differences in boiling points of the different alkane molecules.

▲ **Figure 1** *The fractionating tower*

> **Summary questions**
>
> 1 Describe how crude oil is formed. *(1 mark)*
>
> 2 What is the environmental problem associated with fuels that contain sulfur? *(1 mark)*
>
> 3 In fractional distillation what is a *fraction*? *(1 mark)*
>
> 4 Explain how crude oil is separated into different fractions. *(3 marks)*

12.3 Industrial cracking
Specification reference: 3.3.2

> **Key term**
>
> **Cracking:** A decomposition reaction which involves the breaking of C—C single bonds.

> **Maths skill**
>
> Make sure that cracking equations balance correctly.

What is cracking?

The fractional distillation of crude oil produces a range of fractions with different boiling points. The mixture of fractions obtained often contains a higher proportion of heavier fractions than is needed. In industry chemists crack these heavier fractions in order to make lighter, more useful ones. Cracking results in the formation of shorter chain alkanes and alkenes.

The heptane molecule can be cracked in a number of ways. In each equation note that both an alkane and an alkene are formed.

$$C_7H_{16} \rightarrow C_5H_{12} + C_2H_4$$
$$C_7H_{16} \rightarrow C_4H_{10} + C_3H_6$$
$$C_7H_{16} \rightarrow C_3H_8 + 2C_2H_4$$
$$C_7H_{16} \rightarrow C_5H_{10} + C_2H_4 + H_2$$

The shorter chain alkanes are more useful as fuels than the original alkane as they are more volatile (evaporate more readily) and burn more easily. The alkenes are used as raw materials for making polymers such as poly(ethene).

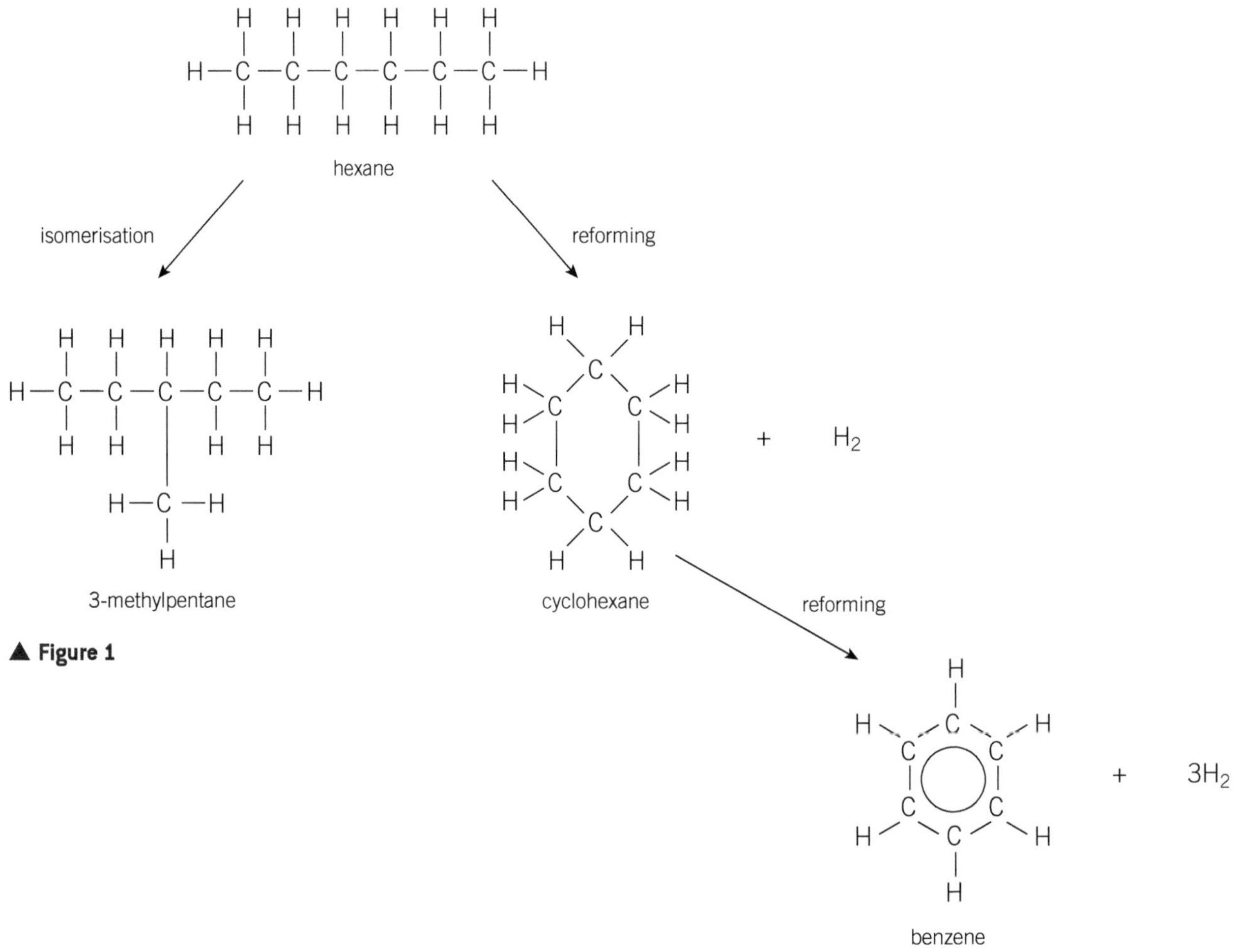

▲ **Figure 1**

Thermal and catalytic cracking

There are two ways cracking can be carried out: thermal cracking and catalytic cracking.

You need to be aware of the differences in reaction conditions used to carry out the two types of cracking.

	Thermal cracking	Catalytic cracking
raw material	long chain alkane	long chain alkane
temperature	800–900 °C	500 °C
pressure	up to 7000 kPa	slightly above atmospheric pressure
catalyst	none	silica and aluminium oxide or zeolite
products	alkene, short chain alkane	aromatic hydrocarbons
uses of products	making polymers	motor fuels (short chains burn more readily, more volatile)
notes	• air is excluded from this process • heating carried out for <1 second to prevent total decomposition.	• more efficient than thermal cracking • produces more branched, cyclic, and aromatic hydrocarbons

Economic reasons for cracking

The fractional distillation process produces disproportionate amounts of some of the fractions of crude oil.

- The amount of petrol produced is not enough for our needs.
- Too much of the naphtha fraction is produced.

> **Revision tip**
> The cracking process helps us make maximum use of crude oil by converting the less useful fractions into more useful ones. This is an essential process, as petroleum is a non-renewable resource.

Summary questions

1. Complete the equations below to show the cracking of some alkane molecules.
   a $C_8H_{18} \rightarrow C_5H_{12} + \underline{}$
   b $C_9H_{20} \rightarrow \underline{} + C_2H_4$
   c $C_{22}H_{46} \rightarrow \underline{} + C_4H_8$ (3 marks)

2. Write equations using displayed formulae for the cracking of:
   a Hexane, C_6H_{14} where one of the products is ethene.
   b Octane, C_8H_{18} where one of the products is propene. (2 marks)

12.4 Combustion of alkanes

Specification reference: 3.3.2

Good fuels

Fuels are substances that can be burnt to release heat energy. Short chain alkanes are good fuels which burn completely in a good supply of air to produce carbon dioxide and water and they release a lot of heat energy. Combustion reactions are always exothermic (have a negative sign).

Types of combustion

Complete combustion

For complete combustion to occur:

- A large excess of oxygen is needed.
- alkane + oxygen → carbon dioxide + water (+ release of energy)

For methane (the main fossil fuel in natural gas):

$$CH_4(g) + 2O_2(g) \rightarrow CO_2(g) + 2H_2O(l)$$

For octane, a longer chain hydrocarbon, the products are identical.

$$C_8H_{18}(g) + 12\tfrac{1}{2}O_2(g) \rightarrow 8CO_2(g) + 9H_2O(l)$$

Incomplete combustion

In a *limited supply* of oxygen, incomplete combustion occurs.

- The hydrogen in the hydrocarbon still forms water.
- The carbon only undergoes partial oxidation to form carbon monoxide.
- In some cases unburnt carbon particles are released as soot during incomplete combustion.

The incomplete combustion of the same fuels as shown above gives:

$$CH_4(g) + 1\tfrac{1}{2}O_2(g) \rightarrow CO(g) + 2H_2O(l)$$

$$C_8H_{18}(g) + 8\tfrac{1}{2}O_2(g) \rightarrow 8CO(g) + 9H_2O(l)$$

- Carbon monoxide is a toxic gas that binds to the haemoglobin in red blood cells and prevents them carrying oxygen.
- Carbon monoxide is difficult to detect because it is colourless, odourless, and tasteless.

Environmental problems

When alkanes are burnt in engines a lot of heat energy is released and high temperatures can be reached. At these high temperatures enough energy is available for the nitrogen and the oxygen in air to react together forming a number of oxides of nitrogen. These are referred to in general as NO_x. One of the gases formed is nitrogen monoxide which can undergo further oxidation in the air to nitrogen dioxide.

$$N_2(g) + O_2(g) \rightarrow 2NO(g)$$

$$2NO(g) + O_2(g) \rightarrow 2NO_2(g)$$

Some of the fuel passes through the car engine without undergoing oxidation. These gases are called unburned hydrocarbons or volatile organic compounds.

In sunlight these unburned hydrocarbons compounds can react with NO_x to form photochemical smog.

Alkanes

Acid rain

Our rain is naturally acidic because it contains dissolved carbon dioxide from the atmosphere. We use the term acid rain for rain that is more acidic than normal. Acid rain may be formed naturally (from volcanic emissions) or by human activity.

Nitrogen dioxide formed in engines forms nitric acid:

$$2NO_2(g) + H_2O(l) + \tfrac{1}{2}O_2(g) \rightarrow 2HNO_3(aq)$$

Fossil fuels contain sulfur, which forms sulfur dioxide during combustion. This then dissolves in water forming sulfurous acid:

$$SO_2(g) + H_2O(l) \rightarrow H_2SO_3(aq)$$

A series of reactions then lead to the formation of sulfuric acid, $H_2SO_4(aq)$. Acid rain can cause chemical weathering and acidification.

Tackling environmental problems

The catalytic converter

A catalytic converter is fitted to the exhaust system of a car and converts harmful gases into less polluting gases. It removes pollutants from the exhaust gases before they are released through the end of the exhaust. The converter contains a honeycomb mesh of two catalysts (platinum–rhodium and platinum–palladium).

At the platinum–rhodium catalyst the NO_x is reduced to nitrogen.

$$2NO(g) \rightarrow N_2(g) + O_2(g)$$

At the platinum–palladium catalyst two oxidation reactions occur. Any carbon monoxide formed is oxidised to carbon dioxide.

$$CO(g) + \tfrac{1}{2}O_2(g) \rightarrow CO_2(g)$$

Unburned hydrocarbons are oxidised to carbon dioxide and water.

Flue-gas desulfurisation

Flue gas is the waste gas from boilers and furnaces that may contain sulfur dioxide. Flue-gas desulfurisation is the process that removes sulfur dioxide from waste gases. It is used in coal-fired power stations, which release large volumes of sulfur dioxide. A number of reactions occur during the flue-gas desulfurisation process. The key equation is that for the reaction of calcium oxide, CaO, with sulfur dioxide gas forming calcium sulfite:

$$CaO(s) + SO_2(g) \rightarrow CaSO_3(s)$$

The calcium sulfite formed is almost insoluble in water and presents a disposal problem. It can, however, be oxidised to make hydrated calcium sulfate which is gypsum, used to make plasterboard:

$$CaSO_3(s) + \tfrac{1}{2}O_2(g) + 2H_2O(l) \rightarrow CaSO_4 \cdot 2H_2O(s)$$

> **Synoptic link**
>
> You learnt about catalysts in Topic 5.3, Catalysts.

Summary questions

1. Define the term *fuel*. *(1 mark)*

2. Define the term *complete combustion*. *(1 mark)*

3. Write balanced equations for the complete combustion of butane. *(2 marks)*

4. Explain how sulfur in fuels can cause environmental problems. *(1 mark)*

12.5 The formation of halogenoalkanes
Specification reference: 3.3.2

> **Key term**
>
> **Free radical:** A species with an unpaired electron.

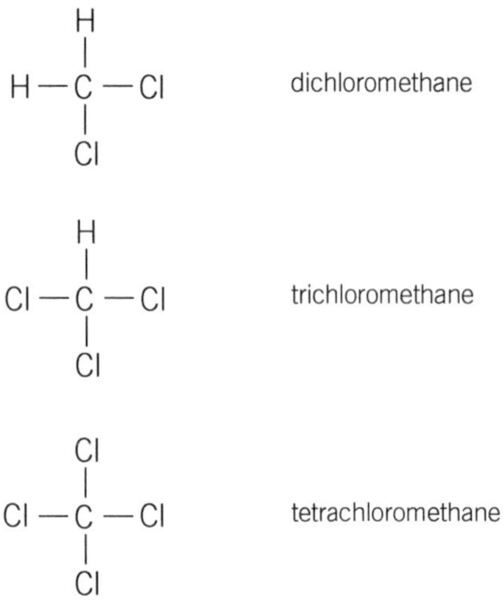

H—C(H)(Cl)—Cl dichloromethane

Cl—C(H)(Cl)—Cl trichloromethane

Cl—C(Cl)(Cl)—Cl tetrachloromethane

▲ **Figure 1** *Products of further substitution reactions*

> **Revision tip**
>
> Many old refrigerators and freezers contain CFCs. When they are disposed of the CFCs must first be removed to prevent the gases escaping into the atmosphere.

> **Summary questions**
>
> 1. Define the term 'free radical'. (1 mark)
>
> 2. Name the type of reaction by which methane reacts with chlorine. (2 marks)
>
> 3. Ethane can react with chlorine to produce chloroethane.
>   a. State the conditions required for the reaction.
>   b. Write an equation for:
>   i. the initiation step
>   ii. the propagation steps
>   iii. a possible termination step.
>   (5 marks)

Free radicals

- Free radicals are very reactive species.
- They are formed when a covalent bond breaks so that one electron is transferred to each of the atoms.
- This is called homolytic fission and forms two free radicals.

$$A–B \rightarrow A\bullet + B\bullet$$

Notice that the dot, which represents the unpaired electron, is written next to the atom that has the unpaired electron.

The free radical substitution of methane

Methane reacts with chlorine by a free radical substitution reaction. You can sum up this reaction using the equation:

$$CH_4 + Cl_2 \rightarrow CH_3Cl + HCl$$

Initiation

- UV light provides the energy required for the homolytic fission of the chlorine molecule.

$$Cl_2 \rightarrow 2Cl\bullet$$

Propagation

- Next the chlorine free radical reacts with methane to form hydrogen chloride and a methyl free radical.

$$Cl\bullet + CH_4 \rightarrow HCl + \bullet CH_3$$

- Then the methyl free radical reacts to form chloromethane and a chlorine free radical.

$$\bullet CH_3 + Cl_2 \rightarrow CH_3Cl + Cl\bullet$$

Notice that as the free radical reacts it forms a new molecule and another free radical.

Termination

In the termination step two free radicals react together to form a new molecule.

$$Cl\bullet + Cl\bullet \rightarrow Cl_2$$
$$\bullet CH_3 + Cl\bullet \rightarrow CH_3Cl$$
$$\bullet CH_3 + \bullet CH_3 \rightarrow C_2H_6$$

Notice that the three steps in the free radical substitution of methane are initiation, propagation, and termination.

- In the initiation step free radicals are made.
- In the propagation steps free radicals react with molecules to form new molecules and free radicals.
- In the termination steps two free radicals join together. A variety of new molecules are formed.

Chapter 12 Practice questions

1. An unbranched saturated hydrocarbon A has the molecular formula C_5H_{12}.
   a. Define the term *saturated hydrocarbon*. (*2 marks*)
   b. Name hydrocarbon A. (*1 mark*)
2. Methane and coal are fuels.
   a. Define the term *fuel*. (*1 mark*)
   b. Give the equation for the complete combustion of methane. (*2 marks*)
   c. Explain how acid rain is formed when fossil fuels, like coal, are burnt. (*2 marks*)
3. The catalytic cracking of hydrocarbon molecules takes place at a lower temperature and pressure than is used in thermal cracking.
   a. Name the catalyst used in the catalytic cracking of hydrocarbons. (*1 mark*)
   b. Explain why this catalyst has a honeycomb structure. (*1 mark*)
4. State and explain the trend in boiling point down the alkane homologous series. (*3 marks*)
5. Propane is a useful fuel.
   a. Give the equation for the complete combustion of propane. (*2 marks*)
   b. Other than carbon dioxide and water vapour name one other product made during the complete combustion of propene. (*1 mark*)
   c. Why does incomplete combustion occur? (*1 mark*)
6. Methane reacts with chlorine to produce chloromethane in a free radical substitution reaction.
   a. Give the overall equation for the reaction between chlorine and methane. (*1 mark*)
   b. Define the term *radical*. (*1 mark*)
   The equation for the initial step is shown below.
   $Cl_2 \rightarrow 2Cl\bullet$
   c. What conditions are required in the initial step? (*1 mark*)
   The propagation step of the free radical substitution reaction takes place in two stages. In the first stage
   $Cl\bullet + CH_4 \rightarrow HCl + CH_3\bullet$
   d. Give the equation for the second step in the propagation step. (*1 mark*)
   e. The termination step can produce ethane. Give the equation for the termination step that produces ethane. (*1 mark*)
7. Which of the following compounds is a member of the alkane homologous series? (*1 mark*)
   a. C_7H_{14}
   b. C_7H_{16}
   c. C_2H_4
   d. CH_2
8. How many atoms are present in one molecule of butane? (*1 mark*)
   a. 14
   b. 12
   c. 9
   d. 16

13.1 Halogenoalkanes – introduction

Specification reference: 3.3.3

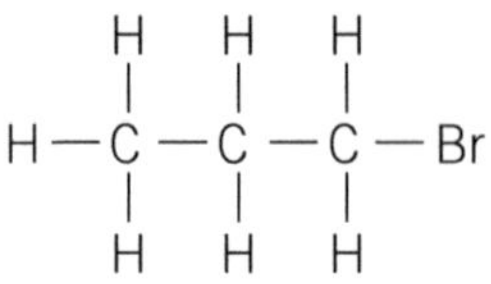

▲ **Figure 1** *1-bromopropane*

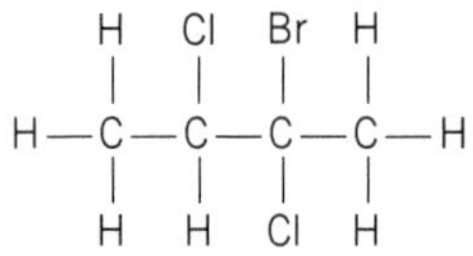

▲ **Figure 2** *2-bromo-2, 3-dichloro butane*

▼ **Table 1** *Table of electronegativity values*

Element	Electronegativity
C	2.5
F	4.0
Cl	3.5
Br	2.8
I	2.6

▼ **Table 2** *Table of bond enthalpy values*

Bond	Bond enthalpy / kJ mol^{-1}
C–F	467
C–H (for comparison)	413
C–Cl	346
C–Br	290
C–I	228

Halogenoalkanes

The halogenoalkanes, $C_nH_{2n+1}X$, are alkanes in which one hydrogen atom has been substituted by a halogen atom, X, where X can be fluorine, chlorine, bromine, or iodine. Their names may include a number to indicate where the halogen atom is bonded onto the carbon chain.

More complicated halogenoalkanes

The prefix di, tri, tetra is sometimes included in names to show the number of each type of halogen atom present. Some compounds contain two different halogen atoms. These halogenoalkanes are named by placing the halogens in alphabetical order.

Why are halogenoalkanes more reactive than alkanes?

Polar bonds

Halogenoalkanes contain a carbon–halogen bond. Halogen atoms are more electronegative than carbon and this results in a polar C–X bond being formed. The greatest difference in electronegativity is in the C–F bond, so the C–F bond is the most polar C–X bond. Bond polarity decreases down the group. These polar bonds mean halogenoalkanes can be attacked by nucleophiles which makes halogenoalkanes more reactive than alkanes.

Strength of the C–X bond

The strength of the C–X bond also influences the reactivity of halogenoalkanes. The strength of the C–X bond decreases down the group. Smaller atoms attract the shared pair of electrons in the C–X bond more strongly so have the greatest bond enthalpy. Down the group the halogen atoms become larger so their attraction for the shared pair of electrons in the C–X bond decreases and the bond becomes weaker.

Table 2 shows that C–Cl, C–Br, and C–I bonds found in chloroalkanes, bromoalkanes, and iodoalkanes are all weaker than the C–H bonds found in alkanes. Many halogenoalkanes are more reactive than alkanes.

Experiments show that the halogenoalkanes become more reactive down the group. This proves that bond enthalpy is a more important factor than bond polarity in predicting the reactivity of halogenoalkanes.

Summary questions

1. Name these halogenoalkanes:
   a, b 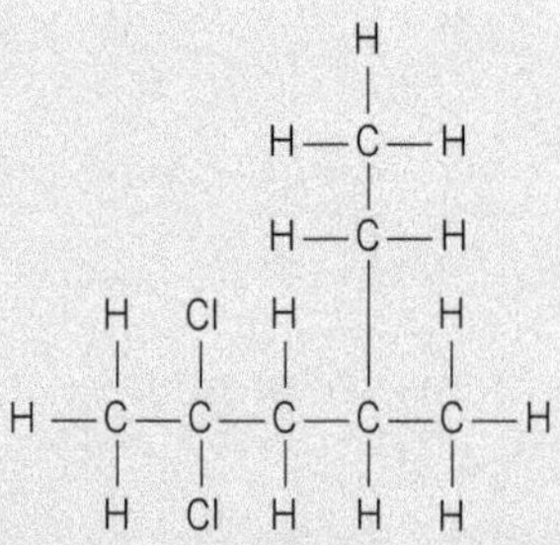 *(2 marks)*

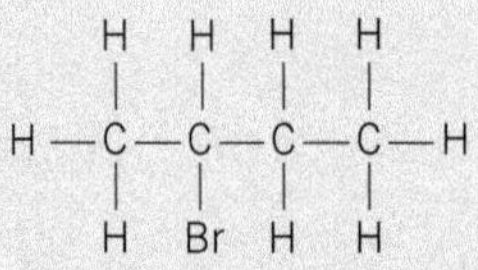

2. State and explain the trend in the bond polarity of the C–X bond down the halogen group. *(2 marks)*

3. State and explain the trend in the bond enthalpy of the C–X bond down the halogen group. *(2 marks)*

13.2 Nucleophilic substitution in halogenoalkanes

Specification reference: 3.3.3

The nucleophilic substitution of halogenoalkanes

In the nucleophilic substitution of a halogenoalkane a nucleophile is added to a molecule and a halide ion is lost.

Hydroxide ions

Hydroxide, OH⁻, ions are nucleophiles which have a lone pair of electrons on the oxygen atom which they can donate to form a new covalent bond. Halogenoalkanes undergo nucleophilic substitution reactions with hydroxide ions (from sodium hydroxide or potassium hydroxide) to form alcohols. Halogenoalkanes do not dissolve in water so ethanol is used as a solvent to allow both reactants to mix and react together. This reaction can also be described as hydrolysis.

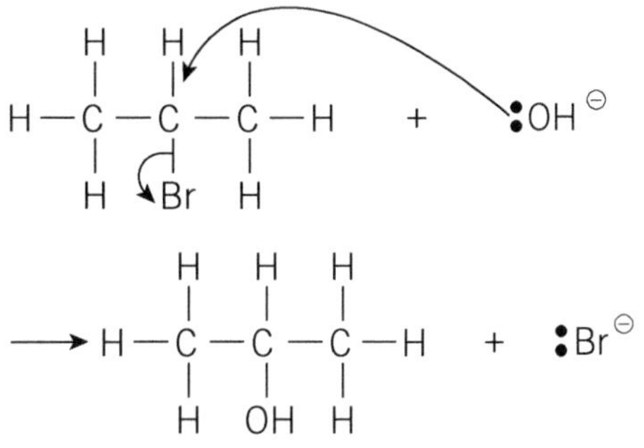

▲ **Figure 1** *Nucleophilic substitution by hydroxide ions*

Cyanide ions

Cyanide, ⁻CN, ions are nucleophiles that have a lone pair of electrons on the nitrogen atom which they can donate to form a new covalent bond. Halogenoalkanes undergo nucleophilic substitution reactions with cyanide, ⁻CN, ions to form nitriles. The halogenoalkane is warmed with potassium cyanide. This reaction is a useful way of increasing the carbon chain length.

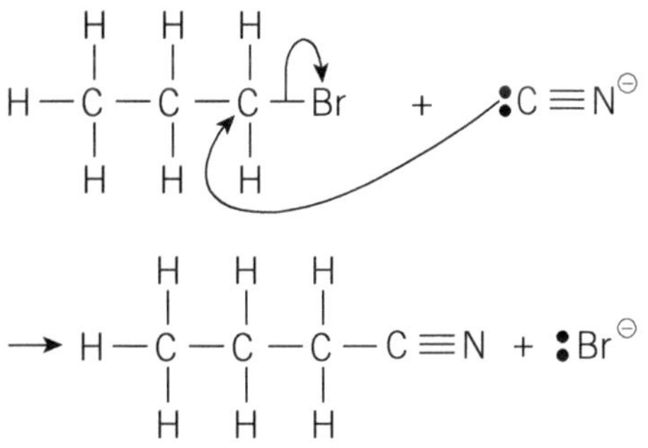

▲ **Figure 2** *Nucleophilic substitution by cyanide ions*

Ammonia molecules

Ammonia, NH_3, molecules are nucleophiles that have a lone pair of electrons on the nitrogen atom which they can donate to form a new covalent bond.

Halogenalkanes undergo nucleophilic substitution reactions with ammonia, NH_3, molecules to form amines. The halogenoalkanes react with concentrated ammonia solution in ethanol at high pressure.

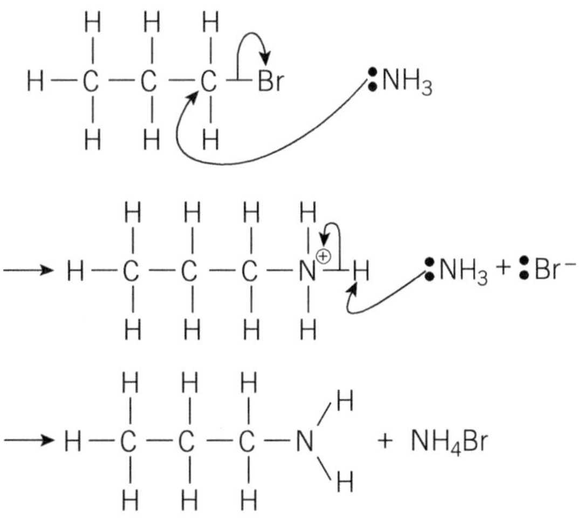

▲ **Figure 3** *Nucleophilic substitution by ammonia molecules*

Using nucleophilic substitution reactions

The nucleophilic substitution reactions of halogenoalkanes are used to introduce new functional groups to organic molecules. In particular using cyanide, ⁻CN, ions allows us to increase the carbon chain length.

Key terms

Reaction mechanism: The route or journey from the reactants to the products through a series of theoretical steps.

Substitution: When a substance swaps one atom or group for another atom or group.

Hydrolysis: When a substance is split up using water molecules or hydroxide ions.

Nucleophiles: Nucleophiles such as ⁻OH, ⁻CN, and NH_3 have a lone pair of electrons which they can donate to another molecule to form a new covalent bond.

Summary questions

1. What is a nucleophile? *(1 mark)*

2. Why is ethanol used in the reaction between 1-bromopropane and sodium hydroxide? *(1 mark)*

3. Give the mechanism for the nucleophilic substitution reaction of 2-chloropropane with cyanide. *(4 marks)*

13.3 Elimination reactions in halogenoalkanes

Specification reference: 3.3.3

Key terms

Primary halogenoalkanes: The halogen is bonded to a carbon which is bonded to just one other carbon atom.

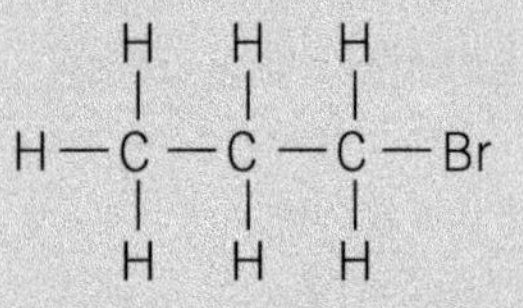

```
    H   H   H
    |   |   |
H — C — C — C — Br
    |   |   |
    H   H   H
```

▲ **Figure 1** *1-bromopropane*

Secondary halogenoalkanes: The halogen is bonded to a carbon which is bonded to two other carbon atoms.

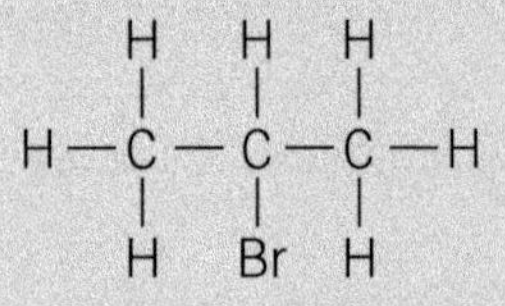

```
    H   H   H
    |   |   |
H — C — C — C — H
    |   |   |
    H   Br  H
```

▲ **Figure 2** *2-bromopropane*

Tertiary halogenoalkanes: The halogen is bonded to a carbon which is bonded to three other carbon atoms.

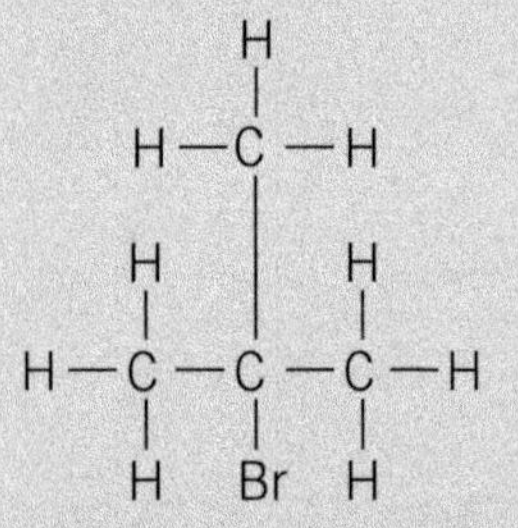

```
        H
        |
    H — C — H
        |
    H   |   H
    |   |   |
H — C — C — C — H
    |   |   |
    H   Br  H
```

▲ **Figure 3** *2-bromo-2-methylpropane*

Elimination reactions

In elimination reactions a small molecule is lost from an organic molecule. When halogenoalkanes undergo elimination reactions a hydrogen halide is lost and an alkene is formed.

The role of hydroxide ions

Hydroxide, OH^-, ions act as nucleophiles in nucleophilic substitution reactions. However under different conditions the hydroxide ions can act as bases. This happens in elimination reaction.

The importance of conditions used

The reaction between halogenoalkanes and hydroxide ions can produce different products depending on the conditions chosen. If the reactants are dissolved in water and the reaction is carried out at room temperature nucleophilic substation reactions are favoured. If the reactants are dissolved in alcohol (ethanol) and the reaction is carried out at high temperature elimination is favoured.

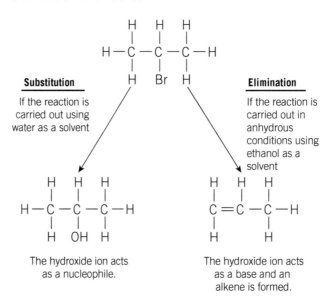

▲ **Figure 4** *The importance of conditions*

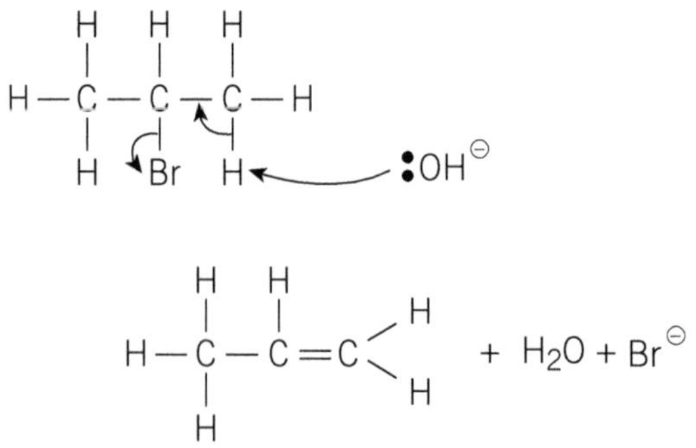

▲ **Figure 5** *Mechanism for the elimination reaction*

Types of halogenoalkane

Primary halogenoalkanes favour nucleophilic substitution reactions while tertiary halogenoalkanes favour elimination reactions. Secondary halogenoalkanes react readily by either mechanism.

Making isomeric alkenes

Elimination reactions can produce a number of different products which are structural and or stereoisomers of each. The reaction below produces three isomers.

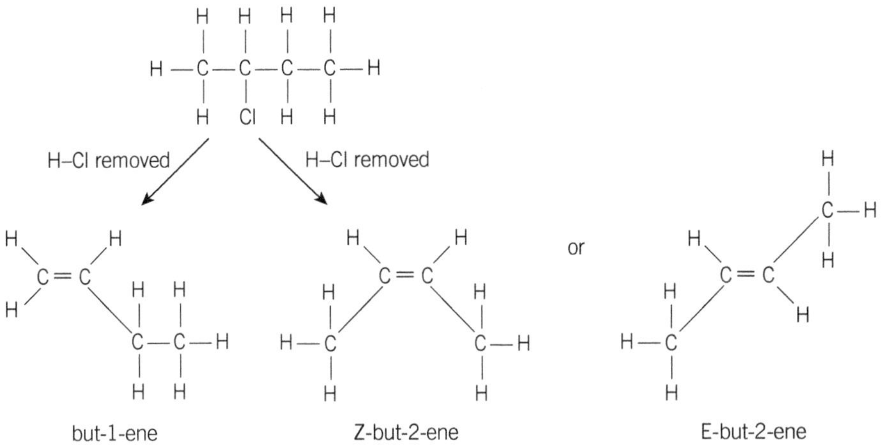

▲ Figure 6

> **Key term**
>
> **Elimination reaction:** In elimination reactions a small molecule, such as water, is lost from an organic molecule.

> **Synoptic link**
>
> Decolourising bromine water is a test for an alkene. The bromine adds on across the double bond. See Topic 14.2, Reactions of alkenes.

> **Key terms**
>
> **Base:** Bases are proton acceptors.
>
> **Structural isomer:** Structural isomers have the same molecular formula but a different structural formula.
>
> **Stereoisomers:** Stereoisomers have the same structural formula but a different arrangement of atoms in space.

Summary questions

1. Define the term *elimination reaction*. *(1 mark)*

2. Name the organic product of the reaction between 2-bromopropane and hydroxide ions in aqueous conditions. *(1 mark)*

3. Give the displayed formula and name the organic products of the reaction between 2-chloropentane and ethanolic potassium hydroxide. *(3 marks)*

Chapter 13 Practice questions

1. Give the skeletal formula and names of the four organic molecules with the molecular formula $C_3H_6Cl_2$. *(4 marks)*

2. A student has a sample of a halogenoalkane which they believe to be a bromoalkane. Outline a method that the student can carry out to confirm that the sample is a bromoalkane. Include any observations that the student should make. *(4 marks)*

3. Consider the two halogenoalkanes below:

    CH_3CH_2Cl

    $CH_3(CH_2)_3Cl$

    a. Name each of these halogenoalkanes. *(2 marks)*

    b. These two halogenoalkanes have different boiling points. State and explain which of these halogenoalkanes has the higher boiling point. *(3 marks)*

4. Bromomethane reacts with an aqueous solution of sodium hydroxide. One of the products is an alcohol. The sodium hydroxide solution contains hydroxide ions which act as nucleophiles.

    a. Define the term *nucleophile*. *(1 mark)*

    b. Complete the diagram to show mechanism for the reaction between bromomethane and hydroxide ions. *(4 marks)*

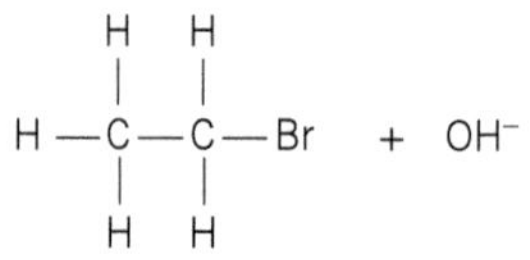

    $$H-\overset{\overset{H}{|}}{\underset{\underset{H}{|}}{C}}-\overset{\overset{H}{|}}{\underset{\underset{H}{|}}{C}}-Br \quad + \quad OH^-$$

    c. Name the reaction mechanism for the reaction between bromomethane and aqueous hydroxide ions. *(2 marks)*

5. Sodium hydroxide is dissolved in ethanol and mixed with 1-chlorobutane. The reaction mixture is then heated. One of the products is an alkene.

    a. Show the reaction mechanism for the reaction that takes place. *(4 marks)*

    b. Name the organic product of the reaction. *(1 mark)*

    c. Name the reaction mechanism for the reaction between bromomethane and ethanolic hydroxide ions. *(4 marks)*

6. Which C–X bond is the most polar?

    A. C–F

    B. C–Cl

    C. C–Br

    D. C–I *(1 mark)*

7. Which C–X bond has the greatest bond enthalpy?

    A. C–F

    B. C–Cl

    C. C–Br

    D. C–I *(1 mark)*

14.1 Alkenes

Specification reference: 3.3.4

Introducing alkenes

Alkenes are unsaturated hydrocarbons with the general formula C_nH_{2n}.

Alkenes have a C=C double bond that consists of a sigma bond and a pi bond. The pi bond forms above and below the axis of the carbon atoms.

Ethene is a planar (flat) molecule because of the C=C arrangement in alkene molecules.

The pi bond has a high electron density (a high chance of the electron being present). Electrophiles can attack this high electron density.

As a result alkenes are more reactive than alkanes.

E–Z stereoisomers

Rotation of the C=C bond would require the pi bond to be broken.

This requires a lot of energy so there is restricted rotation about the C=C double bond.

As a result but-2-ene exists as two stereoisomers.

Notice that in the *E* form the methyl groups are arranged on opposite sides of the C=C bond. The *E* comes from the German word *entgegen* which means opposite.

In the *Z* form the methyl groups are arranged on the same side of the C=C bond. The *Z* comes from the German word *zusammen* which means together.

But-2-ene has stereoisomers because it has two different groups attached to each of the double bonded carbon atoms.

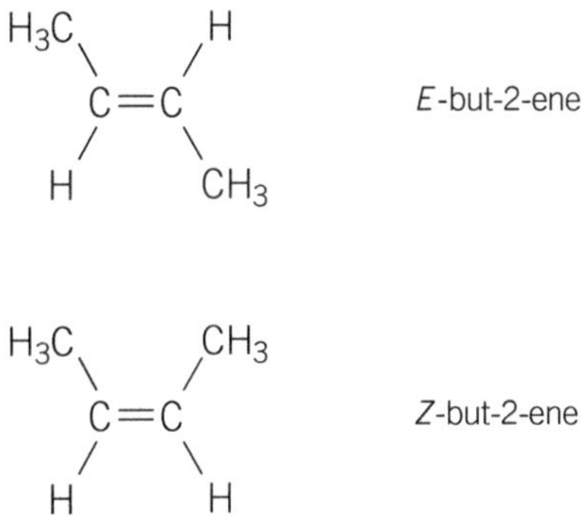

Reactions of alkenes

The bond enthalpy of a C–C bond is 347 kJ mol^{-1} while the bond enthalpy of the C=C bond is 612 kJ mol^{-1}. However, alkenes are much more reactive than alkanes due to the electron-rich C=C bond in the alkene molecules. In addition reactions two species join together. An electrophile is a species that can accept a pair of electrons to form a new covalent bond. Electrophiles can attack the high electron density of the pi bond in alkenes. As a result alkenes undergo electrophilic addition reactions.

Reaction with bromine

Alkenes react with bromine to form dibromoalkanes. For example:

$$C_2H_4 + Br_2 \rightarrow C_2H_4Br_2$$

This is an electrophilic addition reaction.

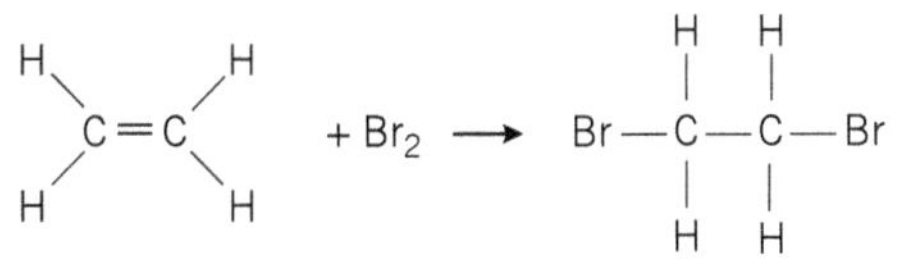

▲ **Figure 2** *The reaction mechanism for the electrophilic addition of bromine to ethene*

Key terms

Unsaturated: An unsaturated hydrocarbon contains at least one carbon double bond.

Electrophile: A species that can accept a lone pair of electrons to form a new covalent bond.

▲ **Figure 1** *E–Z stereoisomers*

Summary questions

1 Define the term *electrophile*. (1 mark)

2 Name the type of reaction that takes place between chlorine and ethene. (1 mark)

3 Draw the displayed formula and name the three alkene molecules which are isomers of each other and have the molecular formula C_4H_8. (3 marks)

14.2 Reactions of alkenes
Specification reference: 3.3.4

Electrophilic addition reactions

The C=C bond in alkene molecules is an electron-rich area that can be attacked by electrophiles. Electrophiles are species than can accept a pair of electrons to form a new covalent bond.

Alkene molecules often undergo electrophilic substitution reactions.

Reaction with hydrogen bromide

Alkenes react with hydrogen bromide to form bromoalkanes. For example:

$$C_2H_4 + HBr \rightarrow C_2H_5Br$$

This is an electrophilic addition reaction.

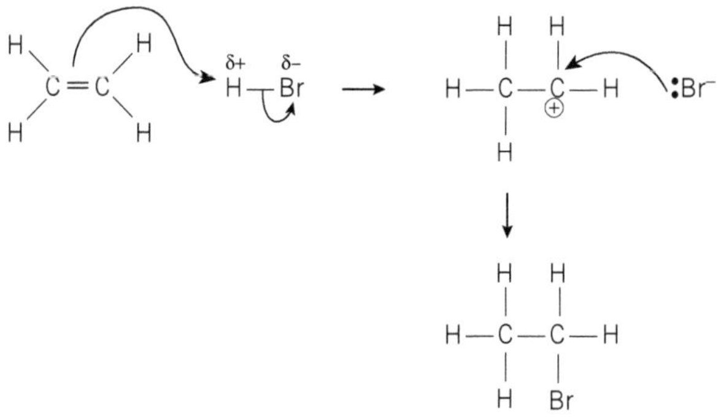

▲ **Figure 1** *The reaction mechanism for the electrophilic addition of hydrogen bromide to ethene*

Revision tip
A useful guide for predicting the major product of an addition reaction with an unsymmetrical alkene is to use Markovnikov's rule.

Markovnikov's rule states that the hydrogen goes to the carbon atom which has the most hydrogens already.

Unsymmetrical alkenes

Ethene is a symmetrical alkene. The groups attached to the C=C bond are the same.

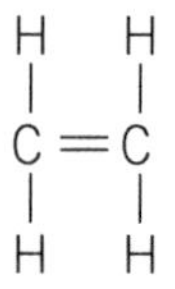

When symmetrical alkenes such as ethene take part in addition reactions with electrophiles such as hydrogen bromide it does not matter how the hydrogen bromide adds across the carbon double bond because the product is always bromoethane.

▲ **Figure 2** *Ethene*

However, this is not true for all alkene molecules.

Propene is an unsymmetrical alkene.

Notice that the groups attached to the carbon atoms in the C=C bond are different. This is significant when propene takes part in an addition reaction with an electrophile such as HBr. The HBr can add across the bond in two different ways, to produce 2-bromopropane or 1-bromopropane.

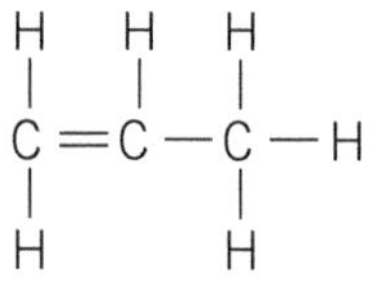

▲ **Figure 3** *Propene*

Summary questions

1. Name the reaction mechanism favoured by alkenes. *(1 mark)*

2. Complete the reaction mechanism below to show the reaction between chlorine and ethene.

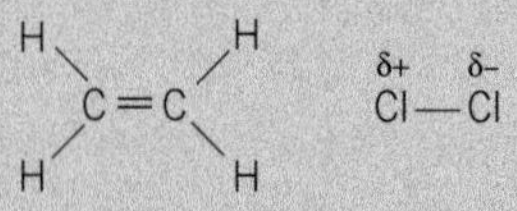

   (4 marks)

3. Draw the displayed formula of the two products of the electrophilic addition of hydrogen chloride to propene. *(2 marks)*

14.3 Addition polymers
Specification reference: 3.3.4

Polymers

Lots of alkene molecules can be joined together to form addition polymers.

- The alkene molecules are called monomers.
- This is an addition polymerisation reaction.

Many ethene molecules can join together to form the addition polymer poly(ethene).

Notice that poly(alkenes) are saturated – they do not have C=C double bonds.

Many propene molecules can join together to form poly(propene).

The section drawn in brackets is called a repeating unit. The repeating unit is repeated thousands of times in each polymer molecule. The lines representing the covalent bonds between repeating units cross through the brackets around the repeating units.

Identifying the monomer

The monomer used to produce an addition polymer can be identified from the repeat unit of the polymer.

First the trailing bonds (bonds through the brackets) are removed.

Then replace the single covalent bond between the carbon atoms with a double covalent bond.

Why are poly(alkenes) unreactive?

- Alkenes have a sigma and a pi bond.
- The high electron density and easy accessibility of the pi bond mean that alkenes can be attacked by electrophiles.
- Poly(alkenes) are saturated hydrocarbons.
  - As a result they are much less reactive than alkenes.

> **Key term**
>
> **Addition polymer:** An addition polymer is formed when lots of small monomers are chemically joined together to form a larger polymer molecule.

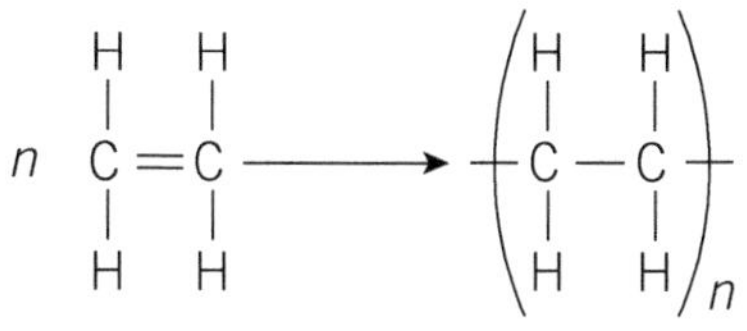

▲ **Figure 1** *The formation of poly(ethene)*

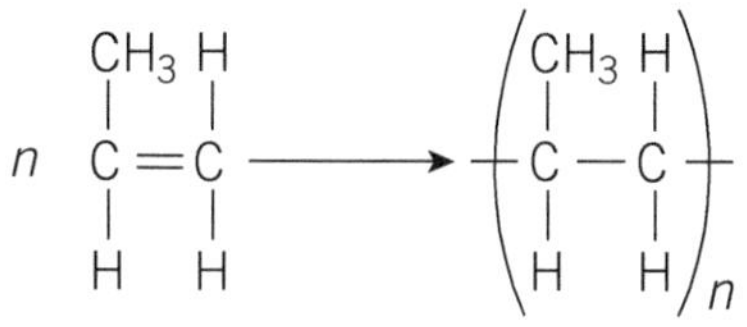

▲ **Figure 2** *The formation of poly(propene)*

> **Common misconception**
>
> Both sulfur dioxide and nitrogen oxides cause acid rain.

Summary questions

1. Name the polymer produced from the addition polymerisation of the following monomers:
   ethene   chloroethene   propene   *(3 marks)*

2. Consider the monomer below.

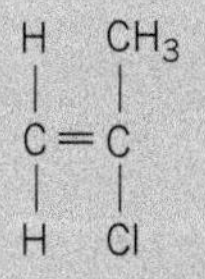

   $$\begin{array}{cc} H & CH_3 \\ | & | \\ C= & C \\ | & | \\ H & Cl \end{array}$$

   Draw the repeat unit of the addition polymer produced from this monomer. *(1 mark)*

3. Consider the repeat unit of the addition polymer shown below. Deduce the monomer used to make this polymer. *(1 mark)*

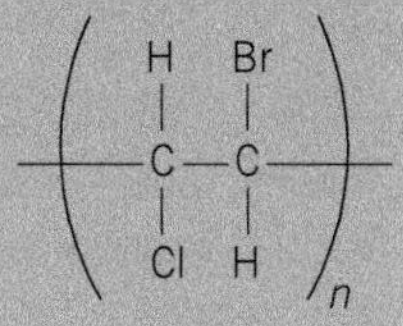

   $$\left(\begin{array}{cc} H & Br \\ | & | \\ -C-C- \\ | & | \\ Cl & H \end{array}\right)_n$$

Chapter 14 Practice questions

1 Ethene reacts with hydrogen chloride to form a new organic compound.
   a Name the new organic compound formed. *(1 mark)*
   b Complete the reaction mechanism for the reaction between ethene and hydrogen chloride. *(5 marks)*

   $$\text{H}_2\text{C}=\text{CH}_2 + \text{H}-\text{Cl}$$

2 A styrene molecule has the displayed formula below.
   a Styrene molecules can be used to make polystyrene.

   Draw the repeat unit of polystyrene. *(1 mark)*
   b Name the type of reaction that produced polystyrene. *(1 mark)*

3 Propene is a fuel.
   a Define the term *fuel*. *(1 mark)*
   b Give the equation for the complete combustion of propene. *(2 marks)*

4 Alkenes are unsaturated hydrocarbons.
   a Define the term *unsaturated hydrocarbon*. *(1 mark)*

   The names and molecular formula of the first five members of the alkene homologous series are shown in the table below.

Name	Molecular formula
ethene	C_2H_4
propene	C_3H_6
butene	C_4H_8
pentene	C_5H_{10}
hexene	C_6H_{12}

   b What is the general formula of alkenes? *(1 mark)*
   c Predict the molecular formula of the seventh member of the alkene homologous series. *(1 mark)*

5 A hydrocarbon is analysed and contains 85.7% carbon by weight. The hydrocarbon has a relative molecular mass of 112.0.
   a Calculate the empirical formula of the hydrocarbon. *(2 marks)*
   b Deduce the molecular formula of the hydrocarbon. *(1 mark)*

6 Give the molecular formula of the monomer that is used to produce polypropene.
   A C_3H_8
   B C_3H_6
   C C_2H_4
   D C_2H_6 *(1 mark)*

7 The test for a carbon double bond is:
   A It turns limewater cloudy.
   B A white precipitate forms when silver nitrate is added.
   C A cream precipitate forms when silver nitrate is added.
   D It decolourises bromine water. *(1 mark)*

15.1 Alcohols – an introduction
Specification reference: 3.3.5

Alcohols
Ethanol is an alcohol.

- Alcohols have the general formula $C_nH_{2n+1}OH$.
- Alcohols contain the –OH or hydroxyl group.
- Alcohols that have more than two carbon atoms can exist as positional isomers, for example, propan-1-ol and propan-2-ol. The number indicates the position of the hydroxyl group.

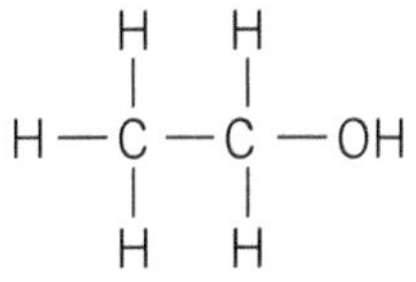

▲ Figure 1 Ethanol

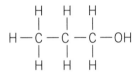

propan-1-ol   propan-2-ol

▲ Figure 2 Positional isomers

Worked example: Draw the structural formula of pentan-2-ol.

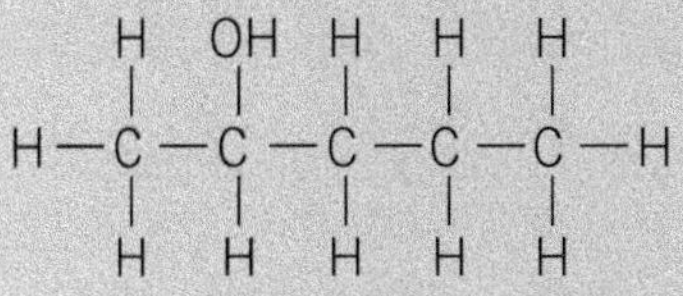

pentan-2-ol

Types of alcohol
We can classify alcohols as primary, secondary, or tertiary depending on the number of carbon atoms attached to the carbon atom bonded to the hydroxyl, –OH, group.

Primary alcohols
In primary alcohols the carbon atom attached to the hydroxyl group is attached to one other carbon atom.

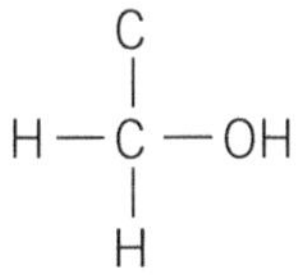

▲ Figure 3 A primary alcohol

Secondary alcohols
In secondary alcohols the carbon atom attached to the hydroxyl group is attached to two other carbon atoms.

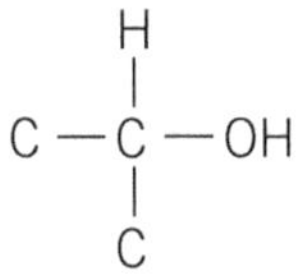

▲ Figure 4 A secondary alcohol

Tertiary alcohols
In tertiary alcohols the carbon atom attached to the hydroxyl group is attached to three other carbon atoms.

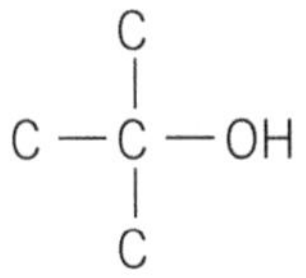

▲ Figure 5 A tertiary alcohol

Worked example: Draw the structural formula of the tertiary alcohol with the formula $C_4H_{10}O$.

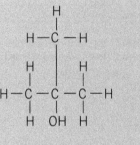

▲ Figure 6 Methylpropan-2-ol

Physical properties of alcohols
Alcohols contain a hydroxyl, –OH, group which means that hydrogen bonds form between alcohol molecules. These strong intermolecular forces of attraction mean that alcohols have relatively high melting points and boiling points compared with alkanes of a similar molecular mass.

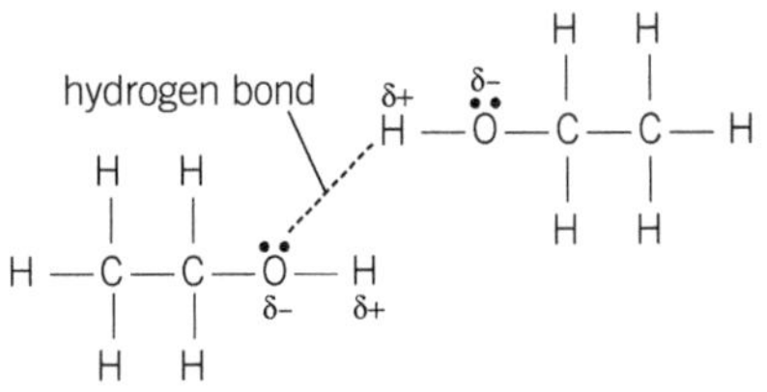

◄ Figure 7 Hydrogen bonding

Alcohols can also form hydrogen bonds to water molecules. This means that short carbon chain length alcohols are soluble in water.

Summary questions
1. What is the general formula of an alcohol? *(1 mark)*
2. What type of alcohol is propan-2-ol? *(1 mark)*
3. Why is methanol soluble in water? Your answer should include a diagram. *(2 marks)*

15.2 Ethanol production
Specification reference: 3.3.5

Producing ethanol
Ethanol can be produced by:
- hydration of ethene
- fermentation.

Hydration of ethene
Ethanol can be made industrially by the hydration of ethene

$$C_2H_4 + H_2O \rightarrow C_2H_5OH$$

This is an addition reaction. Addition reactions have an atom economy of 100%. This is good for sustainable development. All the reactant atoms are made into useful products.

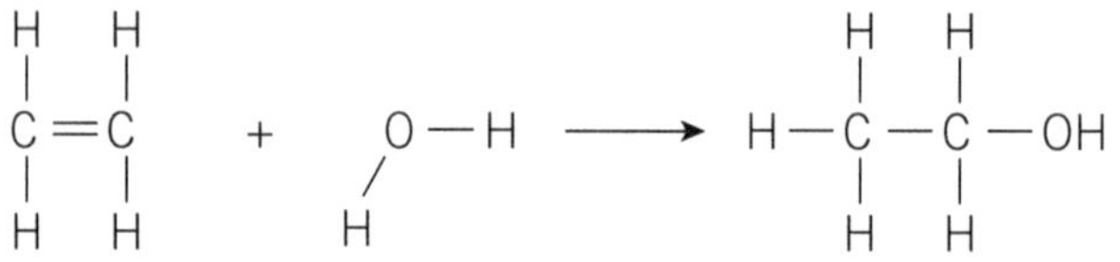

▲ **Figure 1** Producing ethanol

Fermentation
Ethanol can also be made industrially by the fermentation of glucose from plants.

$$\text{glucose} \rightarrow \text{ethanol} + \text{carbon dioxide}$$
$$C_6H_{12}O_6 \rightarrow 2CH_3CH_2OH + 2CO_2$$

Fermentation versus hydration of ethane Biofuels

Fermentation	Hydration of ethene
Slower rate of reaction.	Faster rate of reaction.
Yield of ethanol of around 15%.	Yield of ethanol of around 95%.
Batch process. It is more labour intensive, so labour costs are higher but set up costs are lower.	Continuous process, so labour costs are less but set-up costs are higher.
Atom economy of 51.1%.	An addition reaction with an atom economy of 100%.
Uses lower temperatures and pressures so uses lower amounts of energy.	Uses higher temperatures and pressures so uses higher amounts of energy.
Renewable.	Non-renewable.
Distillation is used to increase the concentration of alcohol. The alcohol that is produced has a lower purity so more steps are required for purification.	Distillation is used to increase the concentration of alcohol. The alcohol produced is already purer so purification is easier.

Summary questions

1 Give two ways that ethanol can be made. *(1 mark)*

2 Which method of making ethanol produces an alcohol that is a renewable? Explain your answer. *(1 mark)*

3 The equation for the hydration of ethene and the fermentation of sugar to make ethanol are shown below.
   a Hydration of ethene
   $C_2H_4 + H_2O \rightarrow C_2H_5OH$
   b Fermentation
   $C_6H_{12}O_6 \rightarrow 2CH_3CH_2OH + 2CO_2$
   Calculate the atom economy of both these methods of making ethanol. Give your answers to two significant figures. *(2 marks)*

15.3 The reactions of alcohols

Specification reference: 3.3.5

Primary and secondary alcohols

Primary and secondary alcohols are oxidised by oxidising agents such as aqueous acidified potassium dichromate(VI), $K_2Cr_2O_7$.

Primary alcohols

Primary alcohols are first oxidised to aldehydes. The aldehyde can be distilled off as it is made to stop any further oxidation.

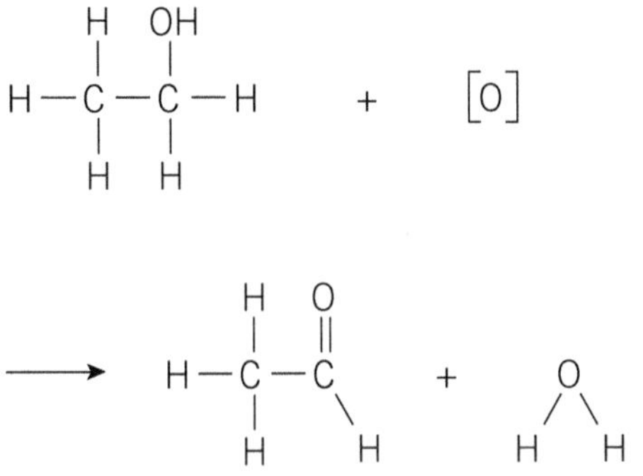

Figure 1 Production of aldehydes

$$CH_3CH_2OH + [O] \rightarrow CH_3CHO + H_2O$$

Notice how the oxidising agent is written as [O] but it is important that the equation still balances.

As the primary alcohols is oxidised the chromium(VI) ion is reduce to chromium(III) and the solution changes colour from orange to green.

If the oxidising agents are in excess and the reaction mixture is heated under reflux then the aldehyde is oxidised to a carboxylic acid.

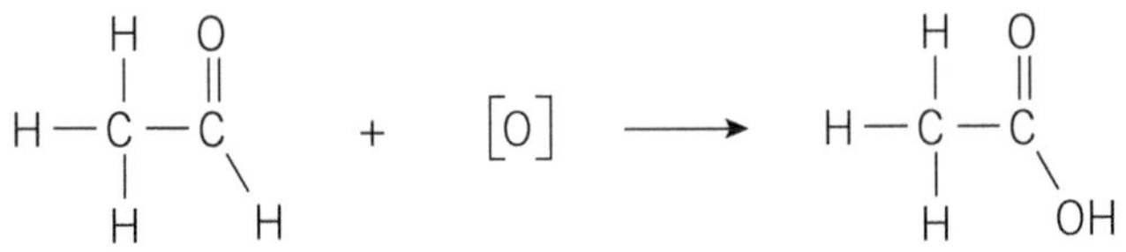

▲ Figure 2 Production of carboxylic acids

$$CH_3CHO + [O] \rightarrow CH_3COOH$$

As the aldehyde is oxidised the chromium(VI) ion is reduced to chromium(III) and the solution changes colour from orange to green.

Secondary alcohols

Secondary alcohols are oxidised to ketones but they cannot be oxidised any further.

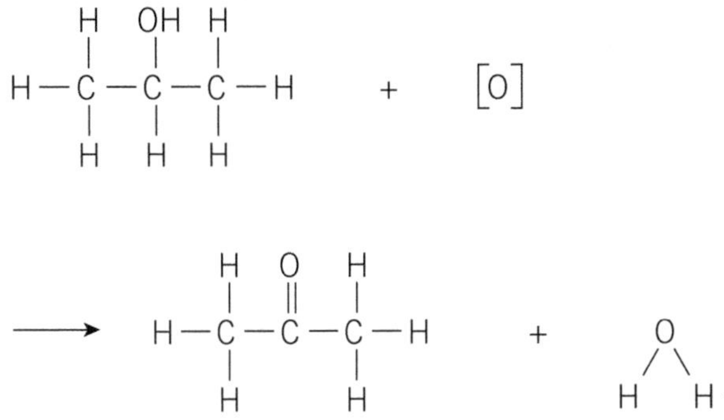

Figure 3 Production of ketones

$$CH_3CH(OH)CH_3 + [O] \rightarrow CH_3COCH_3 + H_2O$$

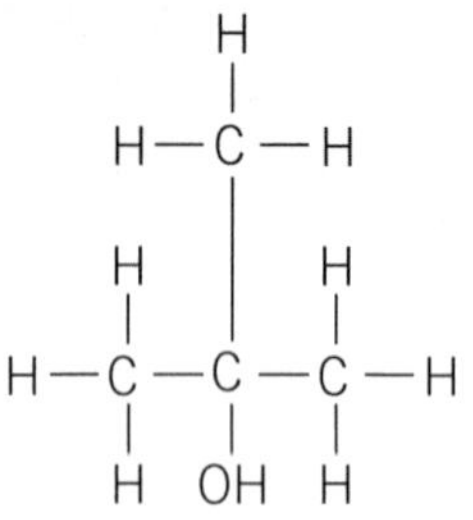

Alcohols

▲ Figure 4 *Methylpropan-2-ol*

Tertiary alcohols

Tertiary alcohols such as methylpropan-2-ol have a hydroxyl group attached to a carbon atom that is attached to three other carbon atoms.

- Tertiary alcohols are not oxidised by oxidising agents such as aqueous acidified potassium dichromate(VI), $K_2Cr_2O_7$. As a result if acidified potassium dichromate(VI) solution is added to a tertiary alcohol no oxidation reaction takes place so the orange solution stays orange.

Distinguishing between aldehydes and ketones

Aldehydes and ketones both contain carbonyl, C=O, groups.

Aldehydes

In aldehydes the carbonyl is at the end of the carbon chain.

Ketones

In ketones the carbonyl is not at the end of the carbon chain.

Tollens' reagent

Tollens' reagent, ammoniacal silver nitrate, is a mild oxidising agent that can be used to distinguish aldehydes from ketones. When Tollens' reagent is added to aldehydes they are oxidised to carboxylic acids and the silver Ag^+ ions in the Tollens' reagent are reduced to silver atoms, Ag, and a silver mirror forms.

Ketones cannot be oxidised so when Tollens' reagent is added to ketones there is no reaction.

Fehling's solution

Fehling's solution is another mild oxidising agent that can be used to distinguish between aldehydres and ketones. It is a blue solution that contains a complex of Cu^{2+} ions.

When Fehling's solution is heated with an aldehyde the aldehyde is oxidised to a carboxylic acid and the Cu^{2+} ions are reduced to Cu^+ ions. A brick-red precipitate of Cu_2O forms.

Ketones cannot be oxidised, so no reaction happens and the blue Fehling's solution does not change colour.

Dehydrating alcohols

Dehydrating agents such as concentrated sulfuric acid or phosphoric acid can be used to remove H_2O from alcohols to form alkenes.

Example

propan-1-ol → propene + water

$CH_3CH_2CH_2OH \rightarrow CH_3CHCH_2 + H_2O$

Notice that as water is lost from the alcohol during the reaction. This may also be classified as an elimination reaction.

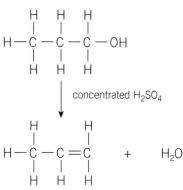

▲ Figure 5 *Production of alkenes*

Key term

Elimination reaction: In elimination reactions a small molecule is lost.

Summary questions

1. What is formed when a secondary alcohol is oxidised? *(1 mark)*

2. Give two reagents that could be used to differentiate between an aldehyde and a ketone. *(2 marks)*

3. Suggest a dehydrating agent that could be used to form an alkene from an alcohol. *(1 mark)*

4. Propan-1-ol is heated with an excess of aqueous acidified potassium dichromate(VI), $K_2Cr_2O_7$. The reaction mixture is heated under reflux.
   a. Name the organic product formed.
   b. Give the equation for the reaction. *(2 marks)*

Chapter 15 Practice questions

1   A sample of an unbranched organic compound was analysed and found to contain only carbon, hydrogen, and oxygen. The sample contained 68.2% carbon and 18.2% oxygen by mass. The relative molecular mass of the organic compound was found to be 88.0.
   a   Calculate the empirical formula of the compound. *(2 marks)*
   b   Deduce the molecular formula of the compound. *(1 mark)*

2   Propan-1-ol can be dehydrated to produce useful new products.
   a   Define the term *dehydration*. *(1 mark)*
   b   Identify a suitable catalyst for the dehydration of propan-1-ol. *(1 mark)*
   c   Name the organic product produced by the dehydration of propan-1-ol. *(1 mark)*

3   Butan-1-ol and 2-methylpropan-2-ol are structural isomers.
   a   What is the molecular formula shared by both compounds? *(1 mark)*
   b   Define the term *structural isomers*. *(1 mark)*
   c   Explain how a student could use a solution of acidified potassium dichromate(VI) to differentiate between these two compounds. *(4 marks)*

4   Primary alcohols can be oxidised by acidified potassium dichromate(VI) to produce useful new organic compounds. The amount of oxidation can be controlled by controlling the conditions used during oxidation. Complete the equations below to show the products of oxidation in each reaction. *(3 marks)*

   a   ethanol structure → [O], alcohol is excess, no reflux → + H_2O

   b   ethanol structure → 2 [O], oxidising agent in excess, reflux → + H_2O

   c   propan-2-ol structure → [O], oxidising agent in excess, reflux → + H_2O

5   Hexanal can be oxidised to a carboxylic acid.
   a   Name the carboxylic acid formed in this reaction. *(1 mark)*
   b   Hexanal can be produced by the oxidation of an alcohol.
      i   Identify this alcohol. *(1 mark)*
      ii State the class to which the alcohol belongs. *(1 mark)*

6   Name the organic product made when butan-2-ol is heated under reflux with acidified potassium dichromate(VI).
   A   butanal
   B   butan-2-one
   C   butan-3-one
   D   butanoic acid *(1 mark)*

16.1 Test-tube reactions

Specification reference: 3.3.6

Synoptic link
Which halogen is present can be determined by the solubility of the precipitate of AgX in ammonia, see Topic 10.3, Reactions of halide ions.

Revision tip
Alkene molecules contain the C=C functional group. Halogenoalkane molecules contain a C-X bond where X is a halogen atom. Alcohols contain the hydroxyl, OH group.

Summary questions

1. How could you confirm that a hydrocarbon was an alkane? *(1 mark)*

2. A student adds an aqueous solution of sodium hydrogencarbonate to a sample of an organic compound. The student observed effervescence. What can the student deduce about the organic compound? *(1 mark)*

3. A sample of an alcohol was heated under reflux with an excess of acidified potassium dichromate(VI) but the reaction mixture stayed orange. What type of alcohol was present? *(1 mark)*

4. A student warms a sample of a carbonyl compound with Tollen's reagent. The student observed that a silver mirror was formed. What can the student deduce about the carbonyl compound being investigated? *(1 mark)*

5. A student has a sample of a halogenoalkane. Outline the steps the student must carry out to confirm that the halogenoalkane is a chloroalkane. *(2 marks)*

Identifying organic compounds

The groups in organic compounds can be identified by some straightforward tests.

Functional group	Test	Result	Notes
Alkenes C=C	Bromine water is added to the sample and the mixture is shaken.	The bromine water decolourises (turns from red-brown to colourless).	Alkanes do not have a C=C bond so do not decolourise bromine water.
Halogenoalkanes R–X	First warm the sample with an aqueous solution of sodium hydroxide. This will hydrolyse the halogenoalkane and produce a halide ion. Then add nitric acid which will remove any impurities. Finally add Tollens' reagent.	A white precipitate of silver chloride shows a chloroalkane. A cream precipitate of silver bromide shows a bromoalkane. A yellow precipitate of silver iodide shows an iodoalkane.	Tollens' reagent is ammoniacal silver nitrate.
Alcohol R–OH	Add acidified potassium dichromate(VI), $K_2Cr_2O_7$ to the sample and warm the reaction mixture.	Primary alcohols are oxidised to form aldehydes and then carboxylic acids. Secondary alcohols are oxidised to form ketones. In both cases a colour change of orange to blue or green is seen.	However, aldehydes (which are not alcohols) would also be oxidised (to carboxylic acids) by acidified potassium dichromate(VI), $K_2Cr_2O_7$ so care must be taken. If the organic substance is oxidised the chromium ions in the potassium dichromate(VI), $K_2Cr_2O_7$ is reduced from +6 (orange) to +4 (blue) or +3 (green).
Aldehyde R–CHO	Add Tolles' reagent Warm the sample with Fehling's solution.	Aldehydes form a silver mirror. A change from a blue solution to a red precipitate	Ketones do not react.
Carboxylic acids R–COOH	Add an aqueous solution of sodium hydrogencarbonate to the organic sample.	Carbon dioxide gas is produced and bubbles are seen.	Carboxylic acids are weak acids and typically have a pH of 3 or 4.

16.2 Mass spectroscopy
Specification reference: 3.3.6

Modern instrumental methods

The mass spectrometer and infrared spectrometer can be used to identify compounds.

These spectrometers can be connected directly to computers, which can process enormous amounts of information at very high speed. Modern instrumental methods can also identify how much of the compound is present.

Molecular ion

The mass spectrometer can be used to measure the relative abundance of the different isotopes of an element. This information can then be used to calculate the relative atomic mass of the element.

Mass spectra can also be used to deduce the relative molecular mass of organic molecules.

The vertical axis of the mass spectrum shows the relative abundance. The horizontal axis of a mass spectrum shows the mass/charge (m/z) ratio of the ions reaching the detector. The ions reaching the detector have a +1 charge. These ions have lost one electron. As electrons have a negligible mass the mass/charge ratio of the heaviest ion which is called the molecular ion reveals the relative molecular mass of the organic molecule.

In Figure 1, the molecular ion has a mass / charge ratio of 75.

The relative molecular mass of propanoic acid is 75.

In the mass spectrum notice how as the organic molecules pass through the mass spectrometer some of the ions fragment into smaller pieces which may then be detected at different mass/charge ratios.

High resolution mass spectroscopy

We use high resolution mass spectroscopy to measure the mass of the molecular ion, M^+, very precisely (to lots of decimal places).

Because of the very high precision used we can determine the relative molecular mass of a compound very precisely and then use this information to deduce the identity of the substance being investigated.

Revision tip
Compared with traditional laboratory techniques, these modern methods of analysis are:
- faster
- more accurate
- more sensitive
- able to use smaller samples.

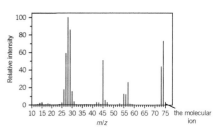

▲ **Figure 1** *The mass spectrum of propanoic acid*

Key term
Molecular ion: The ion formed when a molecule loses a single electron. The molecular ion has a m/z value which is equal to the molecular mass of the molecule.

Worked example: A sample has a relative atomic mass of 30.0688 but is it ethane or methanol?

Element	Relative atomic mass
H	1.0079
C	12.0107
O	15.9994

Ethane, C_2H_6 has a relative atomic mass of

$$2 \times 12.0107 + 6 \times 1.0079 = 30.0688$$

While methanal, CH_2O

$$1 \times 12.0107 + 2 \times 1.0079 + 1 \times 15.9994 = 30.0259$$

So the sample is ethane, C_2H_6

Summary questions

1. What are the advantages of modern methods of analysis compared with traditional methods of analysis? *(1 mark)*

2. What is special about high-resolution mass spectroscopy? *(1 mark)*

3. What does the molecular ion tell us about an organic molecule? *(1 mark)*

16.3 Infrared spectroscopy
Specification reference: 3.3.6

Infrared spectroscopy

Different types of bonds absorb infrared radiation of slightly different wavelength. By seeing which wavelengths have been absorbed we can identify functional groups in organic molecules.

▼ **Table 1** *Identifying bonds in organic molecules*

Typical wavenumber / cm^{-1}	Bond	Location
1000–1300	C–O	alcohols and esters
1620–1680	C=C	alkenes
1680–1750	C=O	aldehydes, ketones, carboxylic acids and esters
2500–3500 (broad)	O–H	hydrogen bonded in carboxylic acids
3230–3550	O–H	hydrogen bonded in alcohols and phenols
3100–3500	N–H	primary amines

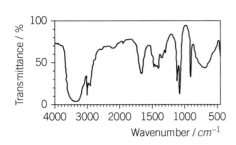

▲ **Figure 1** *Infrared spectrum of ethanol*

Ethanol

Notice that the infrared spectrum of ethanol shows a broad peak at $3400\,cm^{-1}$ caused by the O–H bond and another at $1100\,cm^{-1}$ caused by the C–O bond.

Identifying organic families

Alkenes

Alkenes contain C=C bonds so they will have an absorption band between 1620 and $1680\,cm^{-1}$.

Carbonyls

Aldehydes and ketones both contain C=O bonds so they will have an absorption band between 1650 and $1750\,cm^{-1}$.

Carboxylic acids

Carboxylic acids also contain C=O bonds so they will have an absorption band between 1680 and $1750\,cm^{-1}$. In addition they contain an O–H bond so they will have an absorption band between 2500 and $3500\,cm^{-1}$.

Alcohols

Alcohols contain a C–O bond so they will have an absorption band between 1000 and $1300\,cm^{-1}$. They will also have a O–H bond so they will have a absorption band between 3230 and $3550\,cm^{-1}$.

The fingerprint region

The area of an infrared spectrum between $400\,cm^{-1}$ and $1500\,cm^{-1}$ is known as the fingerprint region because it is unique for every compound. An unknown sample can be identified by comparing its infrared spectrum with a database of known infrared spectra to find a match.

Impurities

Any impurities in a sample will produce absorption bands that should not be there.

Summary questions

1. Would infrared spectroscopy allow a chemist to differentiate between a sample of ethanol and propan-1-ol? Explain your answer. *(1 mark)*

2. How can you use infrared spectroscopy to show that a sample contains impurities? *(1 mark)*

3. A chemist has a sample of an organic compound which has the molecular formula of $C_3H_7O_2$. The infrared spectrum of the compound reveals an absorption band at $1700\,cm^{-1}$ and a very broad absorption band between 2600 and $3500\,cm^{-1}$. Deduce the identity of the organic compound. Explain your answer. *(3 marks)*

Chapter 16 Practice questions

1 A student adds an organic compound containing carbon, hydrogen, and oxygen to a solution of sodium hydrogencarbonate. The student observes bubbles of a gas are produced.
   a Name the gas produced in this reaction. *(1 mark)*
   b Name the class of organic compound present. *(1 mark)*

2 A sample of organic compound A is analysed and is found to be a hydrocarbon.
   a Define the term *hydrocarbon*. *(1 mark)*
   b Describe how a student could confirm that compound A was an alkene. Include the names or any chemicals used and the results of the test in your answer. *(2 marks)*

3 An organic compound A with the molecular formula C_3H_7O was analysed using infrared spectroscopy. The infrared spectrum of compound A showed a peak at $1700\,cm^{-1}$.
   a Which bond was responsible for the peak at $1700\,cm^{-1}$? *(1 mark)*
   b Suggest the names of two possible identities for compound A and explain why it is not possible to be sure of the identity of compound A. *(3 marks)*

4 High resolution mass spectroscopy can be used to find the molecular mass of a parent ion to several decimal places. This can be used to identify the molecular formula of the compound being analysed.

   The table below shows the atomic masses of three elements.

Element	Accurate atomic mass
^{12}C	12.00000
^{1}H	1.007829
^{16}O	15.99491

   Compound X was analysed using high resolution mass spectroscopy and the parent ion was found to have a mass/charge ratio of 200.1049.

   It is suggested that compound X could have a molecular formula

   $C_{11}H_{20}O_3$, $C_{10}H_{16}O_4$, or $C_{11}H_4O_4$

   a Define the term *parent ion*. *(1 mark)*
   b Deduce the molecular formula of compound X. Explain your answer. *(2 marks)*

5 Fehling's solution can be used to differentiate between aldehydes and ketones. Which of these observations would be expected when an aldehyde is warmed with Fehling's solution?
   A blue solution to blue precipitate
   B decolourises
   C blue solution to red precipitate
   D silver mirror *(1 mark)*

6 A chemist analysed a sample of an organic compound using infrared spectroscopy.

   The infrared spectrum shows a peak at $1700\,cm^{-1}$ and a broad peak between 2500 and $3500\,cm^{-1}$. What type of organic compound had been analysed?
   A carboxylic acid
   B alkene
   C alcohol
   D aldehyde *(1 mark)*

17.1 Enthalpy change
17.2 Born–Haber cycles
Specification reference: 3.1.8

Enthalpy change
Definitions for enthalpy changes
You will need to know the following definitions.

Standard conditions are 100 kPa and 298 K.

- **The standard molar enthalpy of formation** $\Delta_f H^\ominus$ is the enthalpy change when one mole of a compound is formed from its constituent elements under standard conditions, all reactants and products in their standard states.

  For example: $H_2(g) + \frac{1}{2}O_2(g) \rightarrow H_2O(l)$ $\Delta_f H^\ominus = -286\,kJ\,mol^{-1}$

  The standard enthalpy of formation of an element is, by definition, zero.

- **The standard enthalpy of atomisation** $\Delta_{at} H^\ominus$ is the enthalpy change which accompanies the formation of one mole of gaseous atoms from the element in its standard state under standard conditions.

  Examples $Mg(s) \rightarrow Mg(g)$ $\Delta_{at} H^\ominus = +147.7\,kJ\,mol^{-1}$

  $\frac{1}{2}Cl_2(g) \rightarrow Cl(g)$ $\Delta_{at} H^\ominus = +121.7\,kJ\,mol^{-1}$

  *This is given per mole of chlorine **atoms** and not per mole of chlorine **molecules**.*

- **First ionisation energy** (first IE) $\Delta_i H^\ominus$ is the standard enthalpy change when one mole of gaseous atoms is converted into a mole of gaseous ions each with a single positive charge.

  For example: $Mg(g) \rightarrow Mg^+(g) + e^-$ $\Delta_i H^\ominus = +738\,kJ\,mol^{-1}$
  so first IE = $+738\,kJ\,mol^{-1}$

- **Second ionisation energy** (second IE) refers to the loss of a mole of electrons from a mole of singly positively charged ions.

  For example: $Mg^+(g) \rightarrow Mg^{2+}(g) + e^-$ $\Delta_i H^\ominus = +1451\,kJ\,mol^{-1}$
  or second IE = $+1451\,kJ\,mol^{-1}$

- **First electron affinity** $\Delta_{ea} H^\ominus$ is the standard enthalpy change when a mole of gaseous atoms is converted to a mole of gaseous ions each with a single negative charge.

  For example: $O(g) + e^- \rightarrow O^-(g)$ $\Delta_{ea} H^\ominus = -141.1\,kJ\,mol^{-1}$
  or first EA = $-141.1\,kJ\,mol^{-1}$

  *This is given per mole of single **atoms**, not per mole of oxygen **molecules** O_2.*

- **Second electron affinity** $\Delta_{ea} H^\ominus$ is the standard enthalpy change when a mole of electrons is added to a mole of gaseous ions each with a single negative charge to form ions each with two negative charges.

  For example: $O^-(g) + e^- \rightarrow O^{2-}(g)$ $\Delta_{ea} H^\ominus = +798\,kJ\,mol^{-1}$ or second electron affinity = $+798\,kJ\,mol^{-1}$

  *This is given per mole of single **atoms**, not per mole of oxygen **molecules** O_2.*

- **Lattice enthalpy of formation** $\Delta_L H^\ominus$ is the standard enthalpy change when one mole of solid ionic compound is formed from its gaseous ions. This is often called the Lattice Energy, LE.

  For example: $Na^+(g) + Cl^-(g) \rightarrow NaCl(s)$ $\Delta_L H^\ominus = -788\,kJ\,mol^{-1}$

> **Key term**
> **Enthalpy change:** Heat change measured at constant pressure.

> **Synoptic link**
> You used Hess's Law in Topic 4.4 to form enthalpy diagrams and enthalpy cycles. You need to know the energy changes for the steps that take place when an ionic compound is formed from its elements.

> **Synoptic link**
> You will need to know the energetics, states of matter, ionic bonding, and change of state studied in Chapter 3 Bonding, and Chapter 4 Energetics.

> **Revision tip**
> Ionisation enthalpies are *always* positive because energy has to be put in to pull an electron away from the attraction of the positively charged nucleus of the atom.

> **Common misconception**
> The second ionisation energy of sodium is *not* the energy change for $Na(g) \rightarrow Na^{2+}(g) + 2e^-$
>
> It is the energy change for $Na^+(g) \rightarrow Na^{2+}(g) + e^-$

> **Revision tip**
> The opposite process, when one mole of ionic compound separates into its gaseous ions, is called **the enthalpy of lattice dissociation** but $\Delta H^\ominus$ is always positive for this process.

When a lattice forms, oppositely charged ions attract together and form a solid. Energy is released. The opposite – separating oppositely charged ions - requires energy.

It is important to have an equation to refer to for enthalpy changes.

Born-Haber cycles

You can work out the lattice enthalpy of formation for an ionic compound using a Born–Haber cycle.

Constructing Born–Haber cycles

Figures 1 and 2 show how the lattice enthalpy of formation is found for a) sodium fluoride and b) calcium sulfide.

- Start the first step from zero
- Positive changes are shown going upwards and negative ones downwards

> **Synoptic link**
> Successive ionisation energies help explain the arrangement of electrons in atoms, see Topic 1.3, The arrangement of electrons.

> **Revision tip**
> First electron affinities are always negative. Second electron affinities are always positive.

> **Key term**
> Born–Haber cycle: Includes all the enthalpy changes for the different steps when an ionic compound is formed from elements in their standard states.

> **Revision tip**
> Hess's Law states that enthalpy change will be the same whatever route you take.

> **Revision tip**
> Remember that the standard enthalpy of atomisation is the enthalpy change for the formation of one mole of gaseous *atoms*. To form one mole of F(g) you start with $\frac{1}{2}F_2(g)$.

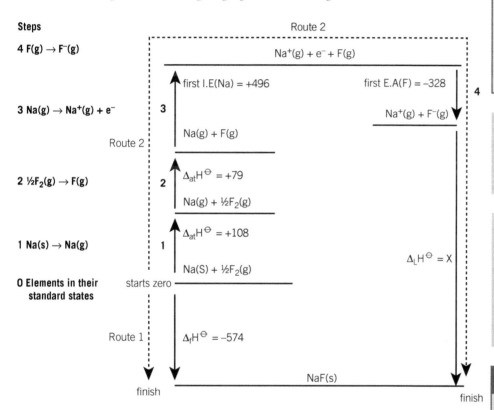

▲ **Figure 1** *Steps in the construction of the Born–Haber cycle to find the lattice enthalpy, X, for sodium fluoride. Enthalpy values are in kJ mol⁻¹. The explanation of each step is given alongside*

The two routes from from $Na(s) + \frac{1}{2}F_2 \rightarrow NaF(s)$ must have the same enthalpy change so:

$-574 = 108 + 79 + 496 - 328 - X$

And $X = 574 + 108 + 79 + 496 - 328 = 929 \text{ kJ mol}^{-1}$

> **Common misconception**
> Lattice enthalpy of formation measures the enthalpy change when *gaseous ions* form the solid compound whereas enthalpy of formation measures the enthalpy change when *elements in their standard state* form the compound.

Thermodynamics

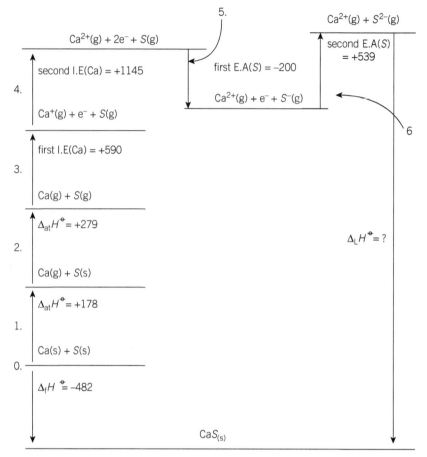

▲ **Figure 2** *Steps in the construction of the Born–Haber cycle to find the lattice enthalpy, Y, for calcium sulfide. Enthalpy values are in kJ mol⁻¹*

Trends in lattice enthalpies

Larger ions mean *smaller* lattice enthalpies – the larger the ions, the greater the distance between the opposite charges of the ions.

For ions of similar radius, the lattice enthalpy increases with size of charge. More energy is released as the ions attract together.

Theoretical compounds

We can predict the lattice energies of compounds theoretically, using the size of the metal ions and the structure of the crystals. Then we can use a Born–Haber cycle to calculate the probable enthalpy of formation.

Summary questions

1. Explain why the first electron affinity of oxygen is exothermic while its second electron affinity is endothermic. *(2 marks)*

2. **a** Write an equation for the third ionisation energy of magnesium. *(1 mark)*
   b The value of the third ionisation energy of magnesium is 7733 kJ mol⁻¹. Explain why it is significantly bigger than the second ionisation energy. *(1 mark)*
   c Calculate $\Delta H^\ominus$ for the process $Mg(g) \rightarrow Mg^{3+}(g) + 3e^-$ *(1 mark)*
   d Calculate the value of $\Delta H^\ominus$ for the process $Mg(s) \rightarrow Mg^+(g) + e^-$ *(2 marks)*

3. State the steps (as in Figure 1) for each numbered step (0 to 6) in Figure 2. *(7 marks)*

4. Calculate the lattice enthalpy of formation for calcium sulfide. *(4 marks)*

5. Explain the 'dip' at the top of the Born–Haber cycle. *(2 marks)*

17.3 More enthalpy changes

Specification reference: 3.1.8

Dissolving ionic compounds

There are three processes *theoretically* involved in order to find $\Delta_{hyd}H^\ominus$ when an ionic solid dissolves in water:

1. The ionic lattice breaks to give separate gaseous ions. The enthalpy change of lattice dissociation, $\Delta_L H^\ominus$ must be put in.
2. The positive ions are hydrated. The enthalpy change of hydration, $\Delta_{hyd}H^\ominus$ is given out.
3. The negative ions are hydrated. The enthalpy change of hydration, $\Delta_{hyd}H^\ominus$ is given out.

Question and model answer: To find $\Delta_{sol}H^\ominus$ (NaCl)

1. $NaCl(s) \rightarrow Na^+(g) + Cl^-(g) \quad \Delta_L H^\ominus = +788 \text{ kJ mol}^{-1}$

   This is the enthalpy change for lattice dissociation.

2. $Na^+(g) + aq + Cl^-(g) \rightarrow Na^+(aq) + Cl^-(g) \quad \Delta_{hyd}H^\ominus = -406 \text{ kJ mol}^{-1}$

   This is the enthalpy change for the hydration of the sodium ion.

3. $Na^+(aq) + Cl^-(g) + aq \rightarrow Na^+(aq) + Cl^-(aq) \quad \Delta_{hyd}H^\ominus = -363 \text{ kJ mol}^{-1}$

   This is the enthalpy change for the hydration of the chloride ion.

So $\Delta_{hyd}H^\ominus(NaCl) = \Delta_L H^\ominus(NaCl) + \Delta_{hyd}H^\ominus(Na^+) + \Delta_{hyd}H^\ominus(Cl^-)$

$+788 - 406 - 363 = +19 \text{ kJ mol}^{-1}$

Key term

Enthalpy of solution $\Delta_{sol}H^\ominus$: The standard enthalpy change when one mole of solute dissolves completely in sufficient solvent to form a solution in which the molecules or ions are far enough apart not to interact with each other.

For example:
$NaCl(s) + aq \rightarrow Na^+(aq) + Cl^-(aq)$
$\Delta_{sol}H^\ominus = +19 \text{ kJ mol}^{-1}$

Key terms

Enthalpy of hydration $\Delta_{hyd}H^\ominus$: of a gaseous ion is the standard enthalpy change when water molecules surround one mole of gaseous ions.

For example:
$Na^+(g) + aq \rightarrow Na^+(aq)$
$\Delta_{hyd}H^\ominus = -406 \text{ kJ mol}^{-1}$

or $Cl^-(g) + aq \rightarrow Cl^-(aq)$
$\Delta_{hyd}H^\ominus = -363 \text{ kJ mol}^{-1}$

Lattice enthalpies and bonding

We can predict the lattice energies of compounds theoretically, using the size of the metal ions and the structure of the crystals. For many ionic compounds, the agreement between the theoretically calculated lattice energy and that measured using a Born–Haber cycle is good. But small highly-charged positive ions can distort the electron density around large highly-charged negative ions. This is called **polarisation** and leads to a degree of covalent bonding, see Figure 1. This will cause a difference between the two values.

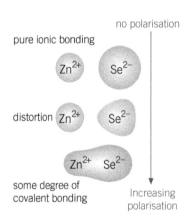

▲ **Figure 1** *Polarisation in zinc selenide*

Summary questions

1. Sodium chloride will dissolve to some extent in methanol, CH_3OH. Suggest which part of the methanol molecule will be attracted to the sodium ions. Explain your answer. *(2 marks)*

2. The lattice energy of potassium bromide is -679 kJ mol^{-1} and the hydration enthalpies of K^+ and Br^- are -351 kJ mol^{-1} and -304 kJ mol^{-1} respectively. Calculate the enthalpy of solution of potassium bromide. *(2 marks)*

3. The enthalpy of hydration of the Li^+ ion is -545 kJ mol^{-1}. Explain why it is much more negative than the value for the sodium ion, Na^+. *(1 mark)*

4. Suggest which of the following mainly ionic compounds will have the greatest covalent character: $AlBr_3$, $MgBr_2$, $NaBr$. Explain your answer. *(2 marks)*

17.4 Why do chemical reactions take place?

Specification reference: 3.1.8

> **Revision tip**
> Entropy and enthalpy have numerical values and units.

There are two driving forces that make chemical reactions happen:

- Increase in **entropy**, S, unit $J\,K^{-1}\,mol^{-1}$ – an increase in randomness.
- Decrease in **enthalpy**, H, unit $kJ\,mol^{-1}$ – a decrease in energy.

Gibbs free energy

Feasible reactions are favoured by *positive entropy changes*, $+\Delta S$, (the products are more random than the reactants) and *negative enthalpy changes*, $-\Delta H$. (energy given out).

These two factors are combined by the Gibbs free energy change, ΔG, given by the equation:

$$\Delta G = \Delta H - T\Delta S$$

> **Key term**
> **Gibbs free energy change:**
> $\Delta G = \Delta H - T\Delta S$

- Feasible reactions have a negative value for ΔG.
- The entropy change has a bigger effect at higher temperature.
- Under some conditions endothermic reactions can be feasible, because the negative value of $T\Delta S$ outweighs the positive value of ΔH.

So to find ΔG we need to know entropy change and enthalpy change for a reaction.

Calculating entropy change, ΔS

You can substitute entropy values into an equation to work out the entropy change ΔS.

> **Revision tip**
> You must divide your final answer for ΔS by 1000 to convert it into $kJ\,K^{-1}\,mol^{-1}$ because ΔH is measured in $kJ\,mol^{-1}$.

> **Worked example: To find ΔS**
>
> For example: $Na(s) + \frac{1}{2}F_2 \rightarrow NaF(s)$
>
> ▼ **Table 1** *Entropies of some substances*
>
> | Substance | State at standard conditions | Entropy $S/\,J\,K^{-1}\,mol^{-1}$ |
> |---|---|---|
> | Na | solid | 51.2 |
> | $\frac{1}{2}F_2$ | gas | 158.6 |
> | NaF | solid | 51.6 |
>
> Entropy of product = $51.6\,J\,K^{-1}\,mol^{-1}$
>
> Entropy of reactants = $51.2 + 158.6 = 209.8\,J\,K^{-1}\,mol^{-1}$
>
> $\Delta S = 51.6 - 209.8 = -158.3\,J\,K^{-1}\,mol^{-1} = -0.1583\,kJ\,K^{-1}\,mol^{-1}$

Is the reaction feasible?

> **Worked example: Finding ΔG**
>
> $\Delta_f H\,NaF(s) = -574\,kJ\,mol^{-1}$
>
> Substitute into $\Delta G = \Delta H - T\Delta S$
>
> $\Delta G = -574 - (298 \times -0.1583) = -574 + 47.173 = -526.8266 = -526.8\,kJ\,mol^{-1}$
>
> So, at room temperature, this reaction is feasible. The reaction is so exothermic that the value of ΔH is much more important than that of ΔS.

Thermodynamics

When ΔG = 0

When $\Delta G = 0$ the reaction is just feasible. So you can calculate the *temperature* at which the reaction just becomes feasible.

Use the relation $0 = \Delta H - T\Delta S$ so that $T = \Delta H/\Delta S$

Worked example: At what temperature will steam decompose to hydrogen and oxygen?

$H_2O(g) \rightarrow H_2(g) + \frac{1}{2}O_2(g)$ $\Delta H = +242\ \text{kJ mol}^{-1}$,
$\Delta S = +45\ \text{J K}^{-1}\text{mol}^{-1} = +0.045\ \text{J K}^{-1}\text{mol}^{-1}$

When $\Delta G = 0$, $T = \Delta H/\Delta S$ so $T = 242/0.045 = 5377$ K

Above this temperature the reaction is feasible.

Revision tip

A negative value of ΔG tells us *nothing* about the reaction rate. For example, ΔG for the reaction of graphite with oxygen is approximately $-400\ \text{kJ mol}^{-1}$, but the reaction is so slow that your pencil does not burst into flames at room temperature.

Summary questions

1. Using the following values of entropy, S: $NH_3(g)$ 192 J K^{-1} mol^{-1}; HCl(g) 187 J K^{-1} mol^{-1}; $NH_4Cl(s)$ 95 J K^{-1} mol^{-1}
   a. Calculate the entropy change, ΔS for the reaction. *(1 mark)*
      $NH_3(g) + HCl(g) \rightarrow NH_4Cl(s)$
   b. Explain the sign of the value you have calculated. *(1 mark)*

2. Predict the approximate entropy change for the reaction.
   $CuO(s) + Zn(s) \rightarrow ZnO(s) + Cu(s)$
   Explain your answer.
   Select from: large and positive, large and negative, approximately zero. *(2 marks)*

3. Using the following values of $\Delta_f G$: $NH_3(g)$ $-16.5\ \text{kJ mol}^{-1}$; HBr(g) $-36.4\ \text{kJ mol}^{-1}$; $NH_4Br(s)$ $-122.4\ \text{kJ mol}^{-1}$
   a. Calculate ΔG for the reaction
      $NH_3(g) + HBr(g) \rightarrow NH_4Br(s)$ *(1 mark)*
   b. What does your answer suggest about the feasibility of the reaction at 298 K? *(1 mark)*
   c. What does your answer suggest about the *rate* of the reaction at 298 K? *(1 mark)*

4. For the reaction $NO(g) + \frac{1}{2}O_2(g) \rightarrow NO_2(g)$ $\Delta H = -56\ \text{kJ mol}^{-1}$ and $\Delta S = -70\ \text{J K}^{-1}\text{mol}^{-1}$
   a. Calculate ΔG for the reaction at 1000 K. *(1 mark)*
   b. Is the reaction feasible at this temperature? Explain. *(2 marks)*

5. Use the data above to calculate ΔG for the decomposition of steam at **a** 1000 K, **b** 10 000 K. State whether the reaction is feasible at each temperature. *(2 marks each)*

Go further: Xenon tetrafluoride

Inert gases *do* form some compounds – one of these is xenon tetrafluoride, which can be formed by direct reaction:

$Xe(g) + 2F_2(g) \rightarrow XeF_4(s)$

The relevant bond enthalpies (bond energies) are:

F—F = 158 kJ mol^{-1}

Xe—F = 130 kJ mol^{-1}

a. Use these bond energies to calculate value for $\Delta_f H$ for XeF_4.

b. The data book value for $\Delta_f H$ for XeF_4 is $-251\ \text{kJ mol}^{-1}$. Explain why there is a discrepancy with the value you have just calculated.

c. What does the sign of $\Delta_f H$ for XeF_4 suggest about the likely stability of XeF_4?

d. What other factor is relevant to the feasibility of the reaction $Xe(g) + 2F_2(g) \rightarrow XeF_4(s)$

e. By looking at the state symbols, suggest the sign of this quantity for the reaction above.

Chapter 17 Practice questions

1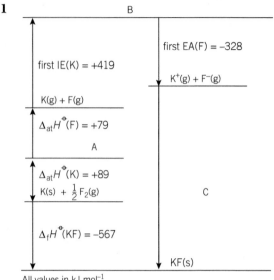

All values in kJ mol^{-1}

    a Fill in the missing areas, A, B, and C in the Born–Haber cycle for potassium fluoride. *(3 marks)*

    b Calculate the lattice energy of potassium fluoride. *(2 marks)*

2 A student measured the entropy change from water to steam using an electric kettle rated at 2.4 kW (2.4 kJ s^{-1}). She let the kettle boil for 100 seconds and found that 100 g of water had boiled away to steam.

At the boiling point, water is equally likely to exist as liquid or gas (there is an equilibrium between liquid and vapour), so ΔG for the change is zero.

    a How many joules of heat were supplied by the kettle in 100 s? *(1 mark)*

    b How many moles of water were boiled away? (M_rs O = 16.0, H = 1.0) *(1 mark)*

    c Rearrange the equation $\Delta G = \Delta H - T\Delta S$ to make ΔS the subject. *(1 mark)*

    d What does the equation become when $\Delta G = 0$? *(1 mark)*

    e What is the boiling point of water in Kelvin? *(1 mark)*

    f Calculate ΔS in J K^{-1} mol^{-1}. *(1 mark)*

    g Explain the sign of ΔS. *(1 mark)*

3 Explain why the lattice enthalpy of magnesium oxide (Mg^{2+}O^{2-}) is approximately four times that of sodium fluoride (Na^{+}F^{-}). *(4 marks)*

4 **a** Given the following values of entropy, calculate ΔS for the reaction
   $MgO(s) + CO_2(g) \rightarrow MgCO_3(s)$
   S / J K^{-1} mol^{-1} MgO(s) 26.9; CO$_2$(g) 213.6; MgCO$_3$(s) 65.7 *(1 mark)*

    b Comment on the sign of your answer. *(1 mark)*

18.1 The rate of chemical reactions
Specification reference: 3.1.9

The rate of a chemical reaction is measured by the change in concentration of one of the products or reactants over time. It is measured in $\text{mol dm}^{-3}\text{s}^{-1}$.

Reaction rate graphs

Figure 1 is a typical concentration/time graph.

Question and model answer: Finding the reaction rate

The tangent to the graph has been drawn at time = 50 s.

Tangents are drawn so they just touch the curve.

The gradient of a tangent is height/base.

Rate of reaction at time
50 s = 0.0035 mol dm⁻³/
120 s = 2.9 × 10⁻⁵ mol dm³ s⁻¹.

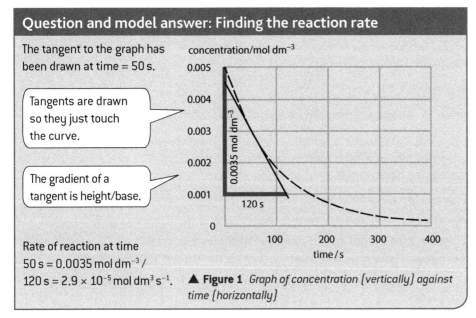

▲ **Figure 1** *Graph of concentration (vertically) against time (horizontally)*

Measuring the reaction rate 🧪

To measure the rate we need to be able to measure the concentration of a reactant or product without disturbing the reaction. Methods include:

- Sampling and titration. Sample the reaction mixture at intervals with a pipette and titrate one of the reactants or products (eg an acid with an alkali). This method is only suitable for relatively slow reactions as the titration takes some time.
- Using a colorimeter – if one of the reactants or products is coloured. The darker the colour, the greater the concentration.
- Collecting a gaseous product.

Revision tip
The reaction rate at any instant is the slope of a tangent drawn on the concentration/time graph at a particular time.

Revision tip
The concentration of a *product* increases with time, the concentration of a *reactant* decreases with time.

Revision tip
The reaction starts fast and gradually slows down as the reactants are used up.

Synoptic link
The initial rate of reaction is measured at the beginning of the reaction when we know exactly what is in the reaction mixture, see Topic 18.3, Determining the rate equation.

Summary questions

1. Look at the reaction rate graph
   a. Is the species whose concentration is being plotted a reactant or a product – explain?
   b. What name is given to the reaction rate represented by the tangent marked A?
   c. What is the reaction rate represented by the tangent marked B?

   (1 mark each)

2. Suggest ways of following each of the following reactions.
   a. $CuSO_4(aq) + Zn(s) \rightarrow ZnSO_4(aq) + Cu(s)$ Hint: $ZnSO_4$ is colourless.
   b. ethanol + ethanoic acid $\rightarrow$ ethyl ethanoate + water
      Hint: the reaction is slow, taking days to go to completion.
   c. $Zn(s) + H_2SO(aq) \rightarrow ZnSO_4(aq) + H_2(g)$ *(1 mark each)*

18.2 The rate expression and order of reaction

Specification reference: 3.1.9

Key term

Rate expression:
Rate = $k[A]^a[B]^b[C]^c$

Revision tip

Remember the use of square brackets to represent concentration in mol dm^{-3}.

Common misconception

Unlike the equilibrium expression (see Topic 6.4, The Equilibrium constant, K_c), the rate expression *cannot* be deduced from the chemical equation for the reaction. It may include species other than the starting materials of the chemical reaction.

Maths skill: Cancelling units

The units of k depend on the overall order.

First order: Rate = $k[A]$, so k = Rate/[A] and the units are ~~mol dm^{-3}~~ s^{-1}/ ~~mol dm^{-3}~~ ie s^{-1}

Second order: Rate = $k[A][B]$, so k = Rate/[A][B] and the units are ~~mol dm^{-3}~~ s^{-1}/ ~~mol dm^{-3}~~ mol dm^{-3} ie s^{-1} mol^{-1} dm^3 or dm^3 mol^{-1} s^{-1}

Third order: Rate = $k[A][B][C]$, k = Rate/[A][B][C] and the units are ~~mol dm^{-3}~~ s^{-1}/ ~~mol dm^{-3}~~ mol dm^{-3} mol dm^{-3} ie s^{-1} mol^{-2} dm^6 or dm^6 mol^{-2} s^{-1}

Work these out rather than try to remember them. You are unlikely to get a reaction of order greater than 3.

The rate expression

The rate equation, also called the rate expression, describes how the rate of a reaction is affected by the concentration of all the species that may be involved in the reaction – not just the reactants, but other species such as catalysts.

It is of the form Rate ∝ $[A]^a[B]^b[C]^c$ or

Rate = $k[A]^a[B]^b[C]^c$

where A, B, and C are species present in the reaction mixture.

k is called the rate constant and is the rate of the reaction when the concentration of all the species involved is 1 mol dm^{-3}. The rate expression can *only* be determined by experiment.

The order of a reaction

The order with respect to a particular species is the power to which the concentration of that species is raised in the rate expression. The overall order is the sum of the orders with respect to all the species in the rate expression.

So if the rate expressions is Rate = $k[X][Y]^2[Z]$, the order with respect to X is 1, with respect to Y is 2, and with respect to Z is 1. The overall order is 1 + 2 + 3 = 4.

Summary questions

1. The reaction
   X(aq) + Y(aq) → Z(aq)
   was found to have a rate expression Rate = $k[X]^2[H^+]$
   a What is:
      i k?
      ii the overall order of the reaction?
      iii the order with respect to X?
      iv the order with respect to Y?
      v the order with respect to H$^+$? *(1 mark each)*
   b Suggest the function of H$^+$. *(1 mark)*
   c What are the units of k? *(1 mark)*

2. a What happens to the rate of a reaction that is first order with respect to X when the concentration of X is **i** doubled, **ii** trebled? *(2 marks)*
   b What happens to the rate of a reaction that is second order with respect to X when the concentration of X is **i** doubled, **ii** trebled? *(2 marks)*
   c What happens to the rate of a reaction that is zero order with respect to X when the concentration of X is **i** doubled, **ii** trebled? *(2 marks)*

18.3 Determining the rate equation
Specification reference: 3.1.9

Finding reaction order
There are two methods for finding the order of a reaction for a particular reactant: a **rate–concentration graph** and the **initial rate method**.

Rate-concentration graphs
Starting with a graph of concentration against time you construct a rate–concentration graph by taking the gradient of the graph at several different concentrations (see Figure 1) and then plot these values against concentration.

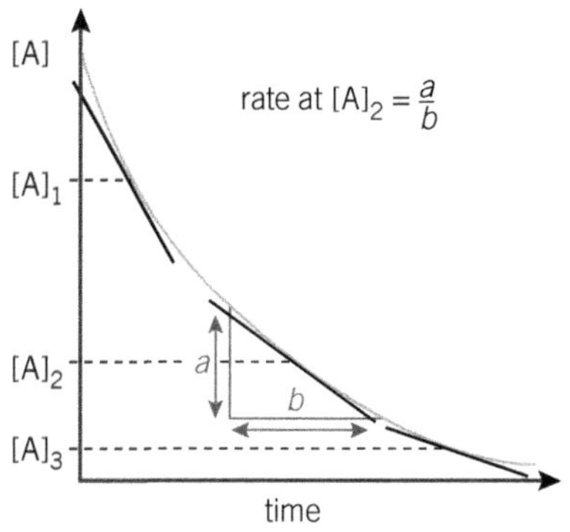

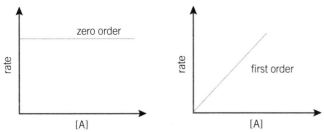

▲ **Figure 1** *Finding the rates of reaction for different values of [A]*

▲ **Figure 2** *Graphs of rate against concentration*

Maths skill: Interpreting graphs
- If the resulting graph is **horizontal straight line** (Figure 2), this means that the concentration of the species being followed does not affect the rate at all — **it is zero order with respect to that species**.
- If the resulting graph is **sloping straight line through the origin** (Figure 2), this means that the rate is directly proportional to the concentration of the species being followed — **it is first order with respect to that species**.
- Any other shape of graph suggests that the order is greater than 1

The initial rate method

Worked example: Using the initial rate method
For the reaction A + B ⟶ products, we do a set of experiments chosen so that pairs of them differ in the concentration of *one* species only. We plot a concentration–time graph for each experiment and draw a tangent to the curve at the start, i.e., t = 0. **This is the initial rate.** The advantage of the initial rate is that we know the concentrations of all the species in the reaction mixture *exactly*. This gives us Table 1.

▼ **Table 1** *Initial rate data*

Experiment No.	[A]	[B]	Initial rate
1	0.5	1.0	2
2	0.5	2.0	8
3	0.5	3.0	18
4	1.0	3.0	36
5	2.0	3.0	72

We now compare the initial rate of pairs of experiments to see how the change of concentration of one species affects the rate. Comparing experiments 3 and 4, [B] is constant and [A] is doubled. The rate doubles, suggesting that the order with respect to A is 1. Comparing experiments 1 and 2, [A] is constant and [B] is doubled. The rate increases by a factor of 4, suggesting that the order with respect to B is 2.

So Rate $\propto$ [A][B]^2 You can confirm this by selecting other suitable pairs.

Summary questions

1. Why is measuring the initial rate of a reaction so useful when determining a rate expression? *(1 mark)*

2. The following initial rates were found for the reaction X + Y ⟶ Z

[X]	[Y]	Initial rate/ mol dm^{-3} s^{-1}
1	1	1
1	2	4
1	3	9
2	3	18
4	3	36

   a What are the orders with respect to X and Y? *(2 marks)*
   b What is the rate expression? *(1 mark)*
   c What are the units of the rate constant? *(1 mark)*
   d Can you be sure that this is the whole rate expression? *(2 marks)*

18.4 The Arrhenius equation
Specification reference: 3.1.9

Synoptic link
The rate constant k is a measure of reaction rate see Topic 18.2, The rate expression and order of reaction.

Revision tip
A very rough rule of thumb says that rates double for a 10°C (10 K) rise in temperature.

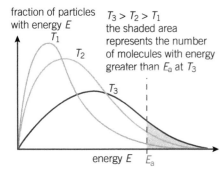

▲ **Figure 1** *Distribution of molecular energies at different temperatures*

Key term
The Arrhenius equation:
$k = {}_A e^{-Ea/RT}$

Revision tip
In calculations with the Arrhenius equation, temperatures must be in Kelvin – add 273 to the temperature in °C.

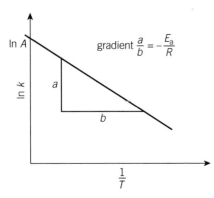

▲ **Figure 2** *An Arrhenius plot*

Maths skill
ln is the symbol for natural log. Make sure you can use your calculator to do natural logs and antilogs.

The effect of temperature on k

Increasing temperature increases the rate of reactions considerably by increasing the number of collisions between reactant molecules and increasing their energy. The latter has by far the greater effect.

To react, molecules must collide with energy greater than the **activation energy, E_a**.

Figure 1 shows the Maxwell–Boltzmann distribution of energies in a gas or liquid. Only the molecules in the shaded area can react.

The shaded area increases rapidly as the temperature increases.

The Arrhenius equation

The Arrhenius equation is derived from the Maxwell–Boltzmann distribution. It is

$k = {}_A e^{-Ea/RT}$

where k is the rate constant, $_A$ is a constant related to the number of molecular collisions, E_a is the activation energy, R is the gas constant (8.3 J K^{-1} mol^{-1}), and T the temperature in Kelvin.

The equation is most useful in its logarithmic form

$\ln k = -E_a/RT + \ln {}_A$

This means that a graph of $\ln k$ (or the log of anything directly proportional to k) against $1/T$ will have a gradient of $-E_a/R$, Figure 2, so we can calculate the activation energy.

Summary questions

1 The graph below shows the Maxwell–Boltzmann distribution of energies at a low temperature.

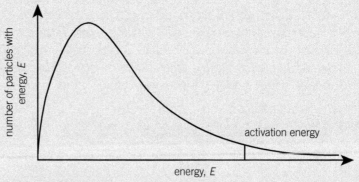

   a How would **i** the height of the peak and **ii** the position of the peak change at a higher temperature? *(2 marks)*
   b What is represented by the area below the whole curve? *(1 mark)*
   c What is represented by the area below the curve to the right of the activation energy line? *(1 mark)*
   d How does the activation energy for the reaction change at a higher temperature? *(1 mark)*

2 In Figure 2 $a = 4.95$ and $b = 0.4 \times 10^{-3}$ K^{-1}
   a Calculate E_a. *(2 marks)*
   b Comment on your answer. *(1 mark)*

18.5 The rate-determining step
Specification reference: 3.1.9

Reaction mechanisms

Many chemical reactions occur in a series of steps involving short-lived **intermediate** species. This series of steps is called the **reaction mechanism**. The reaction mechanism must be consistent with the experimentally-determined rate equation.

The rate-determining step

In a set of reactions which take place one after another, the overall rate of reaction is governed by the rate of the slowest step which is called the **rate-determining step**. Any species which takes part in the mechanism *after* the rate-determining step *cannot* appear in the rate equation.

> **Key term**
>
> **Reaction mechanism:** The series of simple steps by which a reaction occurs.

> **Revision tip**
>
> Remember the rate expression can *only* be determined by experiment and cannot be worked out from the chemical equation.

Question and model answer: The rate-determining step

The reaction $CO + NO_2 \rightarrow CO_2 + NO$ is believed to take place in the following steps

1. $NO_2 + NO_2 \rightarrow NO + NO_3$ slow – the rate-determining step
2. $NO_3 + CO \rightarrow NO_2 + CO_2$ fast

$NO_2 + NO_2 \rightarrow NO + NO_3 \quad NO_3 + CO \rightarrow NO_2 + CO_2$

$NO_2 + \cancel{NO_2} + \cancel{NO_3} + CO \rightarrow NO + \cancel{NO_3} + \cancel{NO_2} + CO_2$

CO appears *after* the slow step so it *cannot* appear in the rate expression. This implies that as soon as some NO_3 is produced it immediately reacts with CO. The actual rate expression has been found by experiment to be **Rate = $k[NO_2]^2$** which is consistent with this mechanism.

> Notice that if you add the two steps together and cancel any species that are on both sides of the arrow you will get the overall equation.

> NO_3 is an intermediate as it is produced in the first reaction and used up in the second.

Summary questions

1. A reaction $A + B \rightarrow C$ takes place in a series of steps
   Step 1 (slow) $A + X \rightarrow W$ Step 2 (fast) $W \rightarrow X$
   Step 3 (fast) $X + B \rightarrow Y$ Step 4 (fast) $Y \rightarrow C$
   a What are the two reactants? *(1 mark)*
   b What is the product? *(1 mark)*
   c Which is the rate-determining step? *(1 mark)*
   d What is the function of X – explain? *(2 marks)*
   e Which species *cannot* appear in the rate expression – explain? *(2 marks)*

2. The reaction
   $2NO + 2H_2 \rightarrow N_2 + 2H_2O$

   is thought to take place in the following steps:
   $NO + NO \rightarrow N_2O_2$ fast
   $N_2O_2 + H_2 \rightarrow N_2O + H_2O$ slow
   $N_2O + H_2 \rightarrow N_2 + H_2O$ fast
   a Add up the steps to show that you get the overall equation. *(1 mark)*
   b State two intermediates. *(2 marks)*
   c Is the rate equation Rate = $k[NO]^2[H_2]$ consistent with this mechanism? *(2 marks)*

Chapter 18 Practice questions

1 The following six experiments were carried out to investigate the kinetics of the reaction

$2H_2(g) + 2NO(g) \rightarrow 2H_2O(g) + N_2(g)$

Experiment No.	Initial [NO] / 10^{-3} mol dm^{-3}	Initial [H$_2$] / 10^{-3} mol dm^{-3}	Initial rate / 10^{-3} mol dm^{-3} s^{-1}
1	6	1	3
2	6	2	6
3	6	3	9
4	1	6	0.5
5	2	6	2
6	3	6	4.5

a What is the order of the reaction with respect to:

   i NO, ii H$_2$, iii overall? Explain your reasoning. *(4 marks)*

b Write the rate expression for the reaction. *(2 marks)*

c Calculate a value for the rate constant and give its units. *(2 marks)*

d Calculate the rate of the reaction when [H$_2$] = 1 × 10^{-3} mol dm^{-3} and [NO] = 1 × 10^{-3} mol dm^{-3}. *(2 marks)*

e Explain why measuring the *initial* rate of a reaction is important. *(1 mark)*

2 The following table gives data about the rate of hydrolysis of a halogenoalkane at different temperatures.

a What is the order of this reaction? Explain your answer. *(2 marks)*

Rate constant, k / 10^{-4} s^{-1}	Temperature / K
0.106	273
3.19	298
9.86	308
29.20	318

b Plot a suitable graph to calculate the activation energy for the reaction. *(10 marks)*

c Is the value you have calculated realistic? Explain your answer. *(2 marks)*

3 The equation for the hydrolysis of bromobutane is

$C_4H_9Br + OH^- \rightarrow C_4H_9OH + Br^-$

What is the overall order of reaction?

  A zero

  B 1

  C 2

  D Impossible to tell. *(1 mark)*

19.1 Equilibrium constant, K_p for homogeneous systems

Specification reference: 3.1.10

Homogeneous systems

In a homogeneous system all the species are in the same phase, for example, all gases or all solutions, as opposed to a heterogeneous system where, for example, one reactant is a solid and another is a gas.

Partial pressure

Partial pressure, p, is a way of expressing concentration for gases.

p of gas A = total pressure × mole fraction

where mole faction = no. of moles of gas A / total no. of moles of gas in the mixture.

The equilibrium law for gases

This is just the same as the equilibrium law expressed in terms of concentration (see Topic 6.4, The Equilibrium constant, K_c), but using partial pressures. So for a reversible reaction

$$aA(g) + bB(g) \rightleftharpoons yY(g) + zZ(g)$$

the equilibrium constant in terms of pressure is

$$K_p = \frac{p^y Y(g)_{eqm} \; p^z Z(g)_{eqm}}{p^a A(g)_{eqm} \; p^b B(g)_{eqm}}$$

The effect of temperature and pressure on gaseous equilibrium

Le Chatelier's principle (see Topic 6.2, Changing the conditions of an equilibrium reaction) applies to gases in the same way as to solutions. **If an equilibrium is disturbed, it moves in the direction that tends to reduce the disturbance.** So increasing the total pressure of a gas will move the equilibrium to the side with fewer molecules, thus reducing the pressure.

The value of K_p changes with temperature, so for an exothermic reaction (heat given out), increasing temperature decreases the value of K_p, that is, the equilibrium moves to the left (reactants are favoured).

> **Key term**
>
> **Partial pressure:** The pressure that each gas in a mixture would exert if all the other gases were removed.

> **Revision tip**
>
> Both concentration and partial pressure are measures of the number of molecules of a particular substance in a given volume.

> **Revision tip**
>
> If the equilibrium has the same number of molecules of gas on both sides of the equation, changing the pressure will have no effect.

> **Summary questions**
>
> 1. What are the partial pressures of each gas in the following mixture at atmospheric pressure, 100 kPa: 4.4 g CO_2, 6.4 g O_2, 5.6 g N_2? Hint: you will need to work out the number of moles of each gas in the mixture. Use M_rs CO_2 44, O_2 32, N_2 28. *(3 marks)*
>
> 2. a Write the expression for K_p for the following gaseous equilibrium. Give the units if all the partial pressures are in Pascals.
> $N_2O_4(g) \rightleftharpoons 2NO_2(g)$ *(1 mark)*
>   b What would be the effect on K_p of increasing the total pressure – explain? *(2 marks)*
>   c What would be the effect on the position of equilibrium of increasing the total pressure – explain? *(2 marks)*

20.1 Electrode potentials and the electrochemical series

Specification reference: 3.1.11

> **Key term**
>
> **Half cell:** A half cell consists of a metal dipped into a solution of a salt of the same metal.

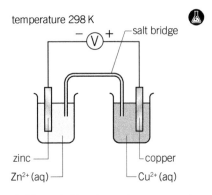

▲ **Figure 1** *Two electrodes connected together with a voltmeter and a salt bridge*

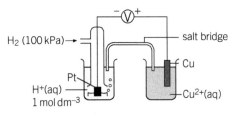

▲ **Figure 2** *Measuring $E^\ominus$ for a copper electrode*

> **Revision tip**
>
> Learn standard conditions: temperature 298 K, concentration of ion in solution 1 mol dm^{-3}, pressure 100 kPa.

▼ **Table 1** *Some values of $E^\ominus$.*

Half reaction	$E^\ominus$/V
Li$^+$(aq) + e$^-$ → Li(s)	−3.03
Ca^{2+}(aq) + 2e$^-$ → Ca(s)	−2.87
Al^{3+}(aq) + 3e$^-$ → Al(s)	−1.66
Zn^{2+}(aq) + 2e$^-$ → Zn(s)	−0.76
Pb^{2+}(aq) + 2e$^-$ → Pb(s)	−0.13
2H$^+$(aq) + 2e$^-$ → H$_2$(g)	0.00
Cu^{2+}(aq) + 2e$^-$ → Cu(s)	+0.34
Ag$^+$(aq) + e$^-$ → Ag(s)	+0.80

Measuring electrode potentials allows us to list metals in order of their reactivity as reducing agents.

Half cells

In a half cell an equilibrium is set up, for example:

Zn(s) ⇌ Zn^{2+}(aq) + 2e$^-$

Electrons build up on the zinc and give it a negative electrical potential.

If two half cells are connected, electrons will tend to flow from the more negatively charged (more reactive) metal to the more positive (less reactive) metal, Figure 1.

The hydrogen electrode

Under standard conditions, the potential of a hydrogen electrode is defined as zero. It is used for measuring the half cell potentials of metals. When a hydrogen electrode is connected to another half cell, the emf recorded is the $E^\ominus$ value for that half cell, see Figure 2.

The electrochemical series

This is a list of half cell potentials with the more negative $E^\ominus$ at the top, see Table 1.

When we connect two half cells, electrons move from negative to positive. If we connect a Zn/Zn^{2+} half cell to a Cu/Cu^{2+} half cell electrons flow from zinc to copper.

Representing cells

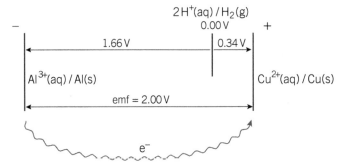

▲ **Figure 3** *Figure 3 Calculating the value of the voltage when two electrodes are connected*

1. Figure 3 shows the shorthand diagram for an Al(s)/Al^{3+}(aq) half cell connected to a Cu(s)/Cu^{2+}(aq) half cell. Electric potential goes from negative on the left to positive on the right of the diagram. Electrons will flow from Al to Cu^{2+}.

2. A second shorthand uses the following conventions:
   - A vertical line shows a **phase boundary**, for example, between solid and solution.
   - A double vertical line shows a salt bridge.

Electrode potentials and electrochemical cells

- The higher oxidation states are written next to the salt bridge.
- Write the emf with the sign of the right hand electrode.

So the cell in Figure 3 would be written

$Al(s)|Al^{3+}(aq) \| Cu^{2+}(aq)|Cu(s)$ $E^\ominus = +2.00\,V$

> **Synoptic link**
>
> All the reactions described in this topic involve electron transfer, so they are **redox reactions**, see Topic 7.1, Oxidation and reduction.

Summary questions

1. **a** What is the emf if a Zn/Zn^{2+} half cell is connected to a Pb/Pb^{2+} half cell? *(1 mark)*
   b Which electrode system would be positive? *(1 mark)*
   c Represent the cell on a conventional cell diagram. *(2 marks)*

2. An electrochemical cell can be represented
   $Al(s)|Al^{3+}(aq) \| Pb^{2+}(aq)|Pb(s)$ $E^\ominus = +1.53\,V$
   a What is represented by the symbols i $|$, ii $\rightleftharpoons$? *(2 marks)*
   b State two errors in the following representation of the cell
   $Al^{3+}(aq)|Al(s) \| Pb(s) | Pb^{2+}(aq)$ $E^\ominus = -1.53\,V$ *(2 marks)*

20.2 Predicting the direction of redox reactions

Specification reference: 3.1.11

We can use diagrams like Figure 3 (in Topic 20.1, Electrode potentials and the electrochemical series) to predict the direction of redox reactions.

Which way will redox equilibria go?

> **Revision tip**
> $E^\ominus$ values are independent of the number of electrons involved.

Question and model answer: Finding the position of an equilibrium

Will the following equilibrium move to the left or the right?

$$2Al(s) + 3Cu^{2+}(aq) \rightleftharpoons 2Al^{3+}(aq) + 3Cu(s)$$

See Figure 3 in Topic 20.1, Electrode potentials and the electrochemical series. Electrons will flow from Al to Cu^{2+} so the half-reactions would be:

$$Al(s) \rightarrow Al^{3+}(aq) + 3e^-$$

$$Cu^{2+}(aq) + 2e^- \rightarrow Cu(s)$$

> In these diagrams, electrons always flow from the half cell on the left to that on the right, that is from the more reactive to less reactive metal in the electrochemical series.

> **Revision tip**
> $E^\ominus$ values only tell you that a reaction is feasible. They say nothing about the rate, which may be so slow that in practice the reaction does not occur.

Multiply the top half reaction by 2 and the second one by 3 to balance the electrons, then add the half-reactions to give

$$2Al(s) + 3Cu^{2+}(aq) \rightarrow 2Al^{3+}(aq) + 3Cu(s)$$

So the equilibrium moves to the right, which is what would happen if aluminium and copper ions were mixed in a test tube.

More complex examples

With no metal and two ions such as $Fe^{3+}(aq)/Fe^{2+}(aq)$ we use a platinum electrode to make contact with a solution containing both the ions at a concentration of 1 mol dm^{-3} to measure $E^\ominus$ as in Figure 1.

▼ **Table 1** $E^\ominus$ values for a range of systems. The best reducing agents are found top right and the best oxidising agents bottom left.

Reduction half equation	$E^\ominus$/V
$Li^+(aq) + e^- \rightarrow Li(s)$	−3.03
$Ca^{2+}(aq) + 2e^- \rightarrow Ca(s)$	−2.87
$Al^{3+}(aq) + 3e^- \rightarrow Al(s)$	−1.66
$Zn^{2+}(aq) + 2e^- \rightarrow Zn(s)$	−0.76
$Cr^{3+}(aq) + e^- \rightarrow Cr^{2+}(aq)$	−0.41
$Pb^{2+}(aq) + 2e^- \rightarrow Pb(s)$	−0.13
$2H^+(aq) + 2e^- \rightarrow H_2(g)$	0.00
$Cu^{2+}(aq) + e^- \rightarrow Cu^+(aq)$	+0.15
$Cu^{2+}(aq) + 2e^- \rightarrow Cu(s)$	+0.34
$I_2(aq) + 2e^- \rightarrow 2I^-(aq)$	+0.54
$Fe^{3+}(aq) + e^- \rightarrow Fe^{2+}(aq)$	+0.77
$Ag^+(aq) + e^- \rightarrow Ag(s)$	+0.79
$Br_2(aq) + 2e^- \rightarrow 2Br^-(aq)$	+1.07
$Cl_2(aq) + 2e^- \rightarrow 2Cl^-(aq)$	+1.36
$MnO_4^- + 8H^+(aq) + 5e^- \rightarrow Mn^{2+}(aq) + 4H_2O(l)$	+1.51
$Ce^{4+}(aq) + e^- \rightarrow Ce^{3+}(aq)$	+1.70

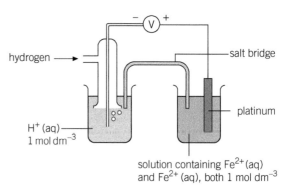

▲ **Figure 1**

The cell diagram would be written as follows

Pt|H_2(aq)|$2H^+$(aq) || Fe^{3+}(aq), Fe^{2+}(aq)|Pt
$E^\ominus = +0.77$ V

The two ions are separated by a comma and the more oxidised ion is written next to the salt bridge.

Electrode potentials and electrochemical cells

Worked example: Will Fe^{3+} oxidise Cl^- to Cl_2?

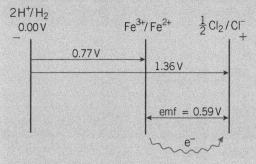

▲ **Figure 2** *Figure 2 Cell diagram for Fe^{3+}/Fe^{2+} connected to $\frac{1}{2}Cl_2/Cl^-$*

Figure 2 shows that electrons will flow from Fe^{3+}/Fe^{2+} to $\frac{1}{2}Cl_2/Cl^-$, ie the Fe^{3+}/Fe^{2+} system releases electrons and $\frac{1}{2}Cl_2/Cl^-$ accepts them. So the half equations are

$Fe^{2+}(aq) \rightarrow Fe^{3+}(aq) + e^-$

$\frac{1}{2}Cl_2(aq) + e^- \rightarrow Cl^-(aq)$

So the full equation is

$Fe^{2+}(aq) + \frac{1}{2}Cl_2(aq) \rightarrow Fe^{3+}(aq) + Cl^-(aq)$

So Fe^{3+} will not oxidise Cl^- to Cl_2: the reverse reaction occurs.

Summary questions

1. Predict whether Fe^{3+} will oxidise Br^- ions to bromine. *(2 marks)*

2. **a** Complete the labelling of the diagram below which shows an experiment to find $E^\ominus$ for a Cu^{2+}/Cu electrode *(7 marks)*

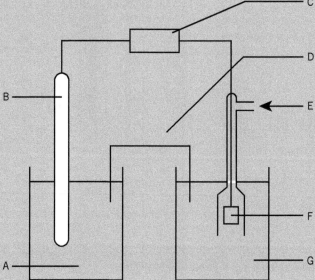

   b At what temperature should the measurement be made? *(1 mark)*

20.3 Electrochemical cells

Specification reference: 3.1.11

Commercial cells are of three types – non-rechargeable (irreversible), rechargeable, or fuel cells.

Non-rechargeable cells

> **Revision tip**
> If we reverse a half cell reaction, we must change the sign of the $E^\ominus$ value.

Worked example: The Daniell cell

The Daniell cell, Figure 1, combines a Zn/Zn^{2+} half cell with a Cu/Cu^{2+} half cell. The copper can form one electrode and the porous pot acts as a salt bridge.

The electrons flow from zinc to copper and the reactions are:

$Zn(s) \rightarrow Zn^{2+}(aq) + 2e^-$ $E^\ominus = -0.76\,V$

$Cu^{2+}(aq) + 2e^- \rightarrow Cu(s)$ $E^\ominus = -0.34\,V$

(The value of $E^\ominus$ is made negative because this is the reverse of the half equation in the electrochemical series.)

Adding the half equations and cancelling the electrons gives the overall reaction.

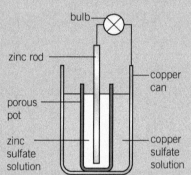

▲ **Figure 1** A Daniell cell

$Zn(s) + Cu^{2+}(aq) \rightarrow Zn^{2+}(aq) + Cu(s)$ $E^\ominus = 1.10\,V$ with zinc as the negative electrode.

The Daniell cell is not very practical because it contains liquid.

Question and model answer: What are the Leclanché cell reactions?

The Leclanché cell uses a paste of ammonium chloride as the electrolyte. The positive electrode is a carbon rod. The reactions are:

$Zn(s) \rightarrow Zn^{2+}(aq) + 2e^-$

$2NH_4^+(aq) + 2e^- \rightarrow 2NH_3(aq) + H_2(g)$

We now add the half reactions

$2NH_4^+(aq) + Zn(s) \rightarrow 2NH_3(aq) + H_2(g) + Zn^{2+}(aq)$

The emf of the cell is approximately 1.5 V (the conditions are not standard).

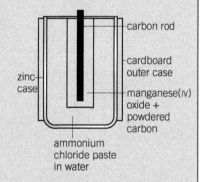

▲ **Figure 2** A Leclanché cell

The carbon rod is surrounded by manganese dioxide which oxidises the hydrogen gas to water to prevent pressure building up inside the sealed cell.

The **zinc chloride cell** is a variant of the Leclanché cell with a different electrolyte paste, while 'long life' batteries use powdered zinc and an electrolyte of potassium hydroxide.

Electrode potentials and electrochemical cells

Rechargeable cells
Lithium ion
The rechargeable lithium ion cell is used in mobile devices. It is light because lithium is the least dense metal.

The positive electrode is made of lithium cobalt oxide, $LiCoO_2$, and the negative electrode is carbon. These are arranged in layers with a sandwich of solid polymer electrolyte in between. On charging, electrons are forced through the external circuit from positive to negative electrode and at the same time lithium ions move through the electrolyte towards the positive electrode to maintain the balance of charge.

The reactions that occur on charging are:

negative electrode: $Li^+ + e^- \rightarrow Li$ $E^\ominus = -3\,V$

positive electrode: $Li^+(CoO_2)^- \rightarrow Li^+ + CoO_2 + e^-$ $E^\ominus = -+1\,V$

On discharging, the processes are reversed so electrons flow from negative to positive. A single cell gives a voltage of between 3.5 V and 4.0 V.

Lead–acid batteries
Car batteries consist of six 2 V cells connected in series to give 12 V. The electrolyte is sulfuric acid solution. The electrodes are a lead plate (negative) and a lead plate coated with lead sulfate (positive). The overall reaction that occurs on discharge is

$PbO_2(s) + 4H^+(aq) + 2SO_4^{2-}(aq) + Pb(s) \rightarrow 2PbSO_4(s) + 2H_2O(l)$ emf $\approx 2\,V$

This reaction is reversed as the battery is charged from the vehicle's alternator.

Fuel cells
The **alkaline hydrogen–oxygen fuel cell** has two platinum electrodes in an electrolyte of sodium hydroxide. At the negative electrode, the reaction is:

$2H_2(g) + 4OH^-(aq) \rightarrow 4H_2O(l) + 4e^-$ $E^\ominus = -0.83\,V$

At the positive electrode:

$O_2(g) + 2H_2O(l) + 4e^- \rightarrow 4OH^-(aq)$ $E^\ominus = +0.40\,V$

Adding and cancelling gives:

$O_2(g) + 4H_2(g) \rightarrow 2H_2O(l)$ emf = 1.23 V

This is the same reaction as burning hydrogen in oxygen but the energy is released as electricity rather than heat. The only product from the cell is pure water.

> **Revision tip**
> When we add half reactions, we add the $E^\ominus$ values taking account of the signs.

> **Revision tip**
> Technically a battery consists of two or more cells connected together.

Electrode potentials and electrochemical cells

Summary questions

1. Complete the labelling of the diagram below of an alkaline fuel cell.
   (6 marks)

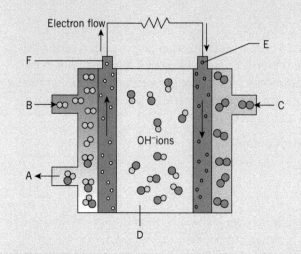

2. Fill in the table with the advantages and disadvantages of different cells.
   (7 marks)

Cell	Advantage	Disadvantage
Daniell	–	
Leclanché		
Lead–acid		
Fuel cell		

3. Explain why hydrogen–oxygen fuel cells are used in manned space craft.
   (2 marks)

Chapter 19 and 20 Practice questions

1. **a** Write the expression for K_p for :
   $$2SO_2(g) + O_2(g) \rightleftharpoons 2SO_3(g)$$
   And give the units of K_p. *(2 marks)*

   b ΔH for this reaction is $-197\,kJ\,mol^{-1}$. State two ways of increasing the yield of SO_3. *(2 marks)*

   c What would be the effect of using a catalyst on **i** the reaction rate
   ii the value of K_p? *(2 marks)*

2. **a** State the units of K_p for the following equilibria:
   i $N_2O_4(g) \rightleftharpoons 2NO_2(g)$
   ii $N_2(g) + O_2(g) \rightleftharpoons 2NO(g)$
   iii $N_2(g) + 3H_2(g) \rightleftharpoons 2NH_3(g)$ *(3 marks)*

   b State the units of K_p for the reverse of the reaction in **a iii**. *(1 mark)*

3. $H_2(g) + I_2(g) \rightleftharpoons 2HI(g)$

   At equilibrium the partial pressures were:

   $H_2(g)$ $2.2 \times 10^4\,Pa$
   $I_2(g)$ $0.5 \times 10^4\,Pa$
   $HI(g)$ $7.3 \times 10^4\,Pa$

   a Calculate the total pressure. *(1 mark)*

   b Calculate the value of K_p and give the correct units. *(2 marks)*

4. **a** Write the conventional cell diagram for a Pb/Pb^{2+} half cell connected to a Cu/Cu^{2+} half cell. Calculate the emf under standard conditions and give the polarity of the copper electrode. Use values of $E^\ominus$ from Table 1 Topic 20.2, Predicting the direction of redox reactions. *(4 marks)*

   b State the standard conditions for this cell. *(2 marks)*

   c Which way will electrons flow if the cell is short circuited? *(1 mark)*

   d Write the half reactions that occur if the cell is short circuited. *(2 marks)*

   e Write the overall reaction that occurs. *(2 marks)*

   f What colour changes would you see if a piece of lead is placed into a solution of copper sulfate? *(2 marks)*

5. Use the $E^\ominus$ values for the half equations below to predict whether bromine solution will oxidise Fe^{2+} ions to Fe^{3+}. Write an equation for any reaction that occurs and calculate the value of the emf of the electrochemical cell formed. *(4 marks)*

   $Fe^{3+}(aq) + e^- \rightarrow Fe^{2+}(aq)$ $E^\ominus = +0.77\,V$
   $Br_2(aq) + 2e^- \rightarrow 2Br^-(aq)$ $E^\ominus = +1.07\,V$

6. State four conditions which must apply if a hydrogen electrode is to have a potential of zero. *(4 marks)*

21.1 Defining an acid

Specification reference: 3.1.12

> **Key term**
>
> **Acid:** A substance that can donate a proton (H^+ ion).
>
> **Base:** A substance that can accept a proton.

Brønsted–Lowry definitions of an acid and a base

These definitions mean that an acid–base reaction need not be in aqueous solution.

For example,

$$HCl(g) + NH_3(g) \rightarrow NH_4^+Cl^-(s)$$

Hydrogen chloride gas has donated a proton to ammonia gas and is therefore an acid.

Ammonia gas has accepted a proton from hydrogen chloride gas and is therefore a base.

Water can act as an acid or a base

For example,

$$HCl(g) + H_2O(l) \rightarrow H_3O^+(aq) + Cl^-(aq)$$

Water is accepting a proton from HCl and is therefore a base in this reaction.

$$NH_3(g) + H_2O(l) \rightarrow OH^-(aq) + NH_4^+(aq)$$

Water is donating a proton to NH_3 and is therefore an acid in this reaction.

The ionic product of water

Water is slightly ionised:

$$H_2O(l) \rightleftharpoons H^+(aq) + OH^-(aq)$$

This equilibrium is set up in water and all aqueous solutions.

> **Revision tip**
>
> This may also be written:
> $2H_2O(l) \rightleftharpoons H_3O^+(aq) + OH^-(aq)$.

You can write an equilibrium expression:

$$K_c = \frac{[H^+(aq)][OH^-(aq)]}{[H_2O(l)]}$$

The concentration of water $[H_2O(l)]$ is constant so we can incorporate it into a new equilibrium constant K_w, where $K_w = K_c \times [H_2O(l)]$.

So, $K_w = [H^+(aq)][OH^-(aq)]$

$$K_w = [H^+(aq)][OH^-(aq)] = 1.0 \times 10^{-14} \text{ mol}^2\text{dm}^{-6}.$$

> **Revision tip**
>
> You must learn this equation. Don't forget the units for K_w.

Every H_2O that dissociates (splits up) produces one H^+ and one OH^- so, in pure water, at 298 K:

$$[OH^-(aq)] = [H^+(aq)]$$

So, $1.0 \times 10^{-14} = [H^+(aq)]^2$

And $[H^+(aq)] = 1.0 \times 10^{-7} \text{ mol dm}^{-3} = [OH^-(aq)]$ at 298 K (25°C)

> **Synoptic link**
>
> K_c, the equilibrium constant, was introduced in Topic 6.4, The Equilibrium constant, K_c.

> **Revision tip**
>
> A base which dissolves in water to produce OH^- ions is called an alkali.

Summary questions

1. Identify the acid and the base in the following equation
$$H_2SO_4 + HNO_3 \rightarrow H_2NO_3^+ + HSO_4^-$$ *(2 marks)*

2. What species are formed after the following acids have donated a proton?
   a HNO_3; **b** HCl; **c** H_2O; **d** HSO_4^- *(4 marks)*

3. In an aqueous alkaline solution, $[OH^-(aq)] = 10^{-2} \text{ mol dm}^{-3}$.
What is $[H^+(aq)]$? *(2 marks)*

21.2 The pH scale
Specification reference: 3.1.12

What is pH?

pH is a measure of the hydrogen ion concentration of a solution, [H$^+$(aq)], using a logarithmic scale.

The acidity of a solution depends on the concentration of hydrogen ions and is measured on the pH scale, where:

pH = –log$_{10}$[H$^+$(aq)]

We use the log scale because [H$^+$] in most aqueous solutions is very small, and the log numbers are much easier to handle.

For example, if [H$^+$(aq)] = 1 × 10^{-14} mol dm^{-3}
$$\log_{10}[H^+(aq)] = -14$$
$$pH = +14$$

Measuring pH

The pH of a solution can be measured using universal indicator paper which usually has a scale from 0 (acid) to 14 (alkali). A pH meter uses an electrode to measure the concentration of H$^+$ ions directly.

Using the pH scale
Finding the concentration of H$^+$(aq) ions

If you know the pH of a solution you can always find [H$^+$(aq)] ions from the equation:

$$pH = -\log_{10}[H^+(aq)]$$

using the antilog (inverse log) function of your calculator.

Finding the concentration of OH$^-$(aq) ions

This takes an extra step.

For example, the pH of a solution is 8.0. What is [OH$^-$(aq)]?

1 Find [H$^+$(aq)] as above.
$$pH = -\log_{10}[H^+(aq)]$$
so $\log_{10}[H^+(aq)] = -8.0$ and [H$^+$(aq)] = 1.0 × 10^{-8} mol dm^{-3}

2 Remember that the ionic product of water is
K_w = [H$^+$(aq)] [OH$^-$(aq)] = 1.0 × 10^{-14} mol^2 dm^{-6}.
Substitute your value for [H$^+$(aq)] = 1.0 × 10^{-8} into this equation:
1.0 × 10^{-8} [OH$^-$(aq)] = 1.0 × 10^{-14} mol^2 dm^{-6}

so [OH$^-$(aq)] = $\frac{1.0 \times 10^{-14}}{1.0 \times 10^{-8}}$ = 1.0 × 10^{-6} mol dm^{-3}

Calculating the pH from [H$^+$(aq)]

Strong acids dissociate *completely* in dilute aqueous solutions.

For example, HCl(aq) → H$^+$(aq) + Cl$^-$(aq)

If we start with hydrochloric acid of concentration 0.5 mol dm^{-3} we know that

[H$^+$(aq)] = 0.5 mol dm^{-3}

pH = –log$_{10}$ [H$^+$(aq)] = –log$_{10}$ 0.5 = +0.30

> **Revision tip**
> Remember that square brackets, [ ], mean the concentration in mol dm^{-3}.

> **Maths skill**
> For example, what is [H$^+$(aq)] if the pH of a solution is 4.5?
>
> 4.5 = –log$_{10}$[H$^+$(aq)] and log$_{10}$[H$^+$(aq)] = –4.5
>
> Look up the antilog of –4.5. It is 3.1622 × 10^{-5}
>
> So [H$^+$(aq)] = 3.2 × 10^{-5} mol dm^{-3} to two significant figures.

> **Revision tip**
> A solution in which [H$^+$(aq)] = [OH$^-$(aq)] is neutral.

> **Summary questions**
>
> 1 Calculate the pH of each of the following solutions:
> a [H$^+$] = 10^{-3} mol dm^{-3};
> b [H$^+$] = 10^{-6} mol dm^{-3};
> c [OH$^-$] = 10^{-4} mol dm^{-3};
> d [H$^+$] = 5 × 10^{-6} mol dm^{-3}
> *(4 marks)*
>
> 2 Assuming sulfuric acid (H$_2$SO$_4$) is fully dissociated in aqueous solution, i.e.,
> H$_2$SO$_4$(aq) → 2H$^+$(aq) + SO$_4^{2-}$(aq),
> what is the pH of a 0.1 mol dm^{-3} solution?
> *(2 marks)*

21.3 Weak acids and bases

Specification reference: 3.1.12

When they are dissolved in water strong acids and bases completely dissociate into ions. Weak acids and bases are only partially ionised in aqueous solution.

Weak acids

Ethanoic acid is the usual example of a weak acid. In a 1 mol dm^{-3} solution the following equilibrium is set up:

$$CH_3COOH(aq) \rightleftharpoons H^+(aq) + CH_3COO^-(aq)$$
ethanoic acid   hydrogen ions   ethanoate ions

Before dissociation: 1000   0   0
At equilibrium: 996   4   4

In this case only 4 in every thousand molecules dissociate. The general equation for the dissociation of a weak acid, HA, is $HA(aq) \rightleftharpoons H^+(aq) + A^-(aq)$

The equilibrium constant for a weak acid is given by:

$$K_a = \frac{[H^+(aq)]_{eqm}[A^-(aq)]_{eqm}}{[HA(aq)]_{eqm}}$$

K_a is called the **acid dissociation constant**. The larger the value of K_a, the stronger the acid, because more of it has dissociated.

Finding the pH of a weak acid

> **Question and model answer: What is the pH of 1.00 mol dm^{-3} ethanoic acid?**
>
> $CH_3COOH(aq) \rightleftharpoons CH_3COO^-(aq) + H^+(aq)$
>
> The concentrations in mol dm^{-3} are:
>
> *First write the equation for the reaction and the expression for K_a.*
>
> $K_a = \frac{[CH_3COO^-(aq)][H^+(aq)]}{[CH_3COOH(aq)]}$
>
> | | $CH_3COOH(aq)$ | $\rightarrow$ | $CH_3COO^-(aq)$ | + | $H^+(aq)$ |
> |---|---|---|---|---|---|
> | Before dissociation: | 1.00 | | 0 | | 0 |
> | At equilibrium: | $1.00 - [CH_3COO^-(aq)]$ | | $[CH_3COO^-(aq)]$ | | $[H^+(aq)]$ |
>
> But as each CH_3COOH molecule that dissociates produces one CH_3COO^- ion and one H^+ ion:
>
> $[CH_3COO^-(aq)] = [H^+(aq)]$
>
> $K_a = \frac{[H^+(aq)]^2}{1.00}$
>
> *The degree of dissociation of ethanoic acid is very small (it is a weak acid), so $[H^+(aq)]_{eqm}$ is so small that $1.00 - [H^+(aq)] \approx 1.00$*
>
> From Table 1, K_a for ethanoic acid
> $= 1.70 \times 10^{-5}$ mol dm^{-3}
>
> So $1.70 \times 10^{-5} = [H^+(aq)]^2$
>
> $[H^+(aq)] = \sqrt{1.70 \times 10^{-5}}$
>
> and $[H^+(aq)] = 4.12 \times 10^{-3}$ mol dm^{-3}
>
> Taking logs: $\log [H^+(aq)] = -2.385$
>
> so pH = 2.385 = 2.39 to 2 d.p

Common misconception

Describing an acid or a base as 'strong or weak' refers to whether an acid or base completely dissociates in water or not. Concentration tells you how much of it is dissolved in water. The two are quite independent.

Table 1 Values of K_a for some weak acids

Acid	K_a / mol dm^{-3}
chloroethanoic	1.30×10^{-3}
benzoic	6.30×10^{-5}
ethanoic	1.70×10^{-5}
hydrocyanic	4.90×10^{-10}

Revision tip
Always quote values of pH to two decimal places.

Revision tip
The pK_a of a weak acid is $-\log_{10} K_a$

Summary questions

1. For propanoic acid $K_a = 1.3 \times 10^{-5}$ mol dm^{-3}
   a Calculate the pH solutions of the following concentrations:
      i 1 mol dm^{-3}
      ii 0.1 mol dm^{-3}
      iii 0.5 mol dm^{-3} *(3 marks)*
   b What is the pK_a of propanoic acid? *(1 mark)*

2. A weak acid HA dissociates in solution
   $HA(aq) \rightarrow H^+(aq) + A^-(aq)$
   What can you say about $[H^+(aq)]$ and $[A^-(aq)]$? *(1 mark)*
   How do the concentrations of $H^+(aq)$ and $HA(aq)$ compare? *(1 mark)*

21.4 Acid–base titrations

Specification reference: 3.1.12

During a titration, the pH of the solution in the flask can be followed with a pH meter, see Figure 1.

pH curves for acid–base titrations

The curves in Figure 2 are obtained by adding a base to a **monoprotic** acid (for example HCl).

> **Revision tip**
> Titrations are introduced in Topic 2.5, Balanced equations and related calculations.

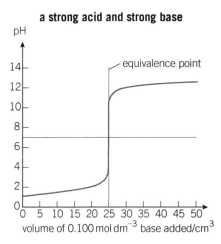
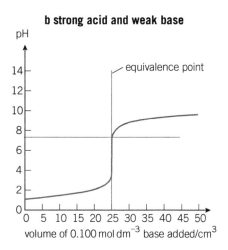
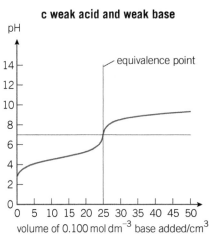
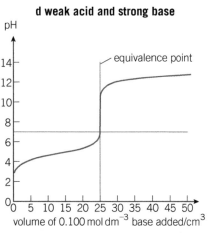

▲ **Figure 2** *Graphs of pH changes for titrations of different acids and bases*

- Each curve has almost horizontal sections where a lot of base can be added without changing the pH much.
- There is also a very steep portion of each curve, except weak acid–weak base, where a single drop of base changes the pH by several units.

The equivalence point

In a titration, the **equivalence point** is the point at which sufficient base has been added to just react with all the acid (or vice versa). In each of the titrations in Figure 2 the equivalence point is reached after 25.0 cm³ of base has been added. Note that the pH at the equivalence point is not always exactly 7.

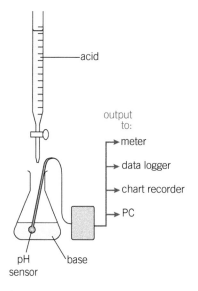

▲ **Figure 1** *Apparatus to investigate pH changes using a pH meter*

Summary questions

1 The curves below show the titration of two different acids A and B with sodium hydroxide.

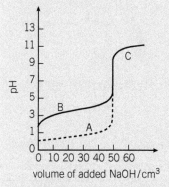

   a Which acid is the strong acid? *(1 mark)*
   b Why are the two curves the same in the region marked C? *(1 mark)*
   c i What is the name of the point when exactly 50 cm³ of sodium hydroxide have been added?
     ii What is the significance of this point? *(2 marks)*

115

21.5 Choice of indicators for titrations

Specification reference: 3.1.12

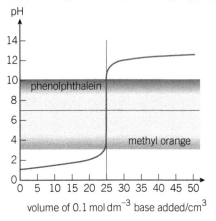
▲ Figure 1

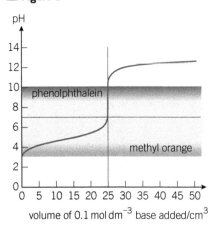
▲ Figure 2

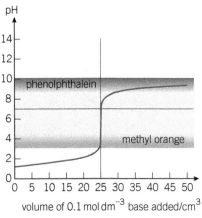
▲ Figure 3

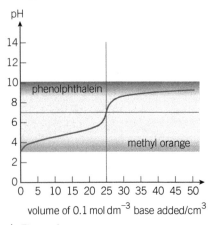
▲ Figure 4

Choosing an indicator

When you do an acid–base titration, your chosen indicator must tell you when there are the same number of hydrogen ions as hydroxide ions in your flask. This point is called the **equivalence point**. The **end point** of a titration is when one drop of added acid (or base) changes the colour of the indicator. These may not be the same unless you choose the correct indicator for your titration.

Properties of a suitable indicator

The end point and the equivalence point must be the same. The colour change must be sharp at the end point. The colour change must be easy to recognise.

Titration curves

▼ Table 1 Some common indicators

Indicator	pK_a value	Colour change
methyl orange	3.7	red — change — yellow
phenolphthalein	9.3	colourless — change — red

pH scale: 0 (very acidic) — 7 (neutral) — 14 (very alkaline)

Figures 1–4 are pH curves for titrations of different combinations of weak and strong acids and bases. In each of these cases the indicator must change at the equivalence point when 25 cm³ of base are added to 25 cm³ of acid.

Figure 1 **Strong acid and strong base**. We need the indicator change to lie within the vertical portion of the pH curve. Both these indicators change when exactly 25 cm³ of base has been added so both would work.

Figure 2 **Weak acid and strong base**. Methyl orange changes colour too soon after the base is added, so phenolphthalein is the right choice here.

Figure 3 **Strong acid and weak base**. Phenolphthalein changes colour too late, so methyl orange is the right choice here.

Figure 4 **Weak acid and weak base**. Neither indicator will work.

The half-neutralisation point

On every titration curve there is always a very gently sloping, almost horizontal gradient on the left-hand side. As you add acid (or base), there is very little change to the pH, almost up to the volume of the equivalence point.

- The point halfway between the zero and the equivalence point is the **half-neutralisation point**. So you know you can add acid or base up to here and the pH will hardly change.
- The pH at the half neutralisation point is the pK_a of a weak acid.

Summary questions

1 Explain why universal indicator is unsuitable for any titration. *(2 marks)*

2 Bromophenol blue changes colour between pH 4.5 and 5.5. State which of the types of titration above (Figures 1–4) it would be suitable for. *(2 marks)*

21.6 Buffer solutions
Specification reference: 3.1.12

How buffers work

Buffers keep the concentration of hydrogen ions and hydroxide ions in a solution almost unchanged. They create an equilibrium reaction so that either additional hydrogen ions or hydroxide ions are removed if these are added.

Acidic buffers

An acidic buffer is made from a mixture of a weak acid and a soluble salt of that acid. It will maintain a pH of below 7 (acidic).

A weak acid, HA, dissociates in solution:
$HA(aq) \rightleftharpoons H^+(aq) + A^-(aq)$

We add to this enough of a soluble salt containing A^- so that there is a similar concentration of A^- ions and HA.

- If we add H^+ ions, Le Chatelier's principle says that the equilibrium moves to the left, removing the added acid.
- If we add OH^- ions, they will combine with the H^+ ions to form water ($H^+(aq) + OH^-(aq) \rightarrow H_2O(l)$) and the equilibrium will move to the right – more HA will dissociate to restore the H^+ ion concentration.

So both added H^+ and OH^- can be removed.

Buffers don't ensure that *no* change in pH occurs. The addition of acid or alkali will still change the pH, but only slightly. It is possible to saturate a buffer by adding so much acid or alkali that all of the available HA or A^- is used up.

Another way of making a buffer is by neutralising a weak acid with an alkali such as sodium hydroxide. If you neutralise half the acid (see Topic 21.5, Choice of indicators for titrations) you end up with a buffer whose pH is equal to the pK_a of the acid, as the solution contains an equal supply of HA and A^-.

At half-neutralisation: pH = pK_a

Basic buffers

Basic buffers also resist change but maintain a pH at above 7. They are made from a mixture of a weak base and a soluble salt of that base, for example, a mixture of aqueous ammonia and ammonium chloride, $NH_4^+Cl^-$.

In this case:

- The aqueous ammonia removes added H^+:
  $NH_3(aq) + H^+(aq) \rightarrow NH_4^+(aq)$
- The ammonium ion, NH_4^+, removes added OH^-:
  $NH_4^+(aq) + OH^-(aq) \rightarrow NH_3(aq) + H_2O(l)$

Some examples of buffers

Blood is buffered so that the pH is maintained at approximately 7.4. A change of as little as 0.5 of a pH unit may be fatal. Detergents and shampoos are buffered. If they become too acidic or too alkaline, they could damage fabric or skin and hair. Buffers are important in brewing – the enzymes that control many of the reactions involved work best at specific pH values.

> **Key term**
>
> **Buffers:** Solutions that can resist changes of pH when small amounts of acid or alkali are added to them.

> **Synoptic link**
>
> Le Chatelier's principle is covered in Topic 6.2, Changing the conditions of an equilibrium reaction.

Acids, bases, and buffers

Calculating the pH of a buffer

Buffers can be made which will maintain different pHs.

When a weak acid dissociates:
$$HA(aq) \rightleftharpoons H^+(aq) + A^-(aq)$$

$$K_a = \frac{[H^+(aq)][A^-(aq)]}{[HA(aq)]}$$

You can use this expression to calculate the pH of buffers.

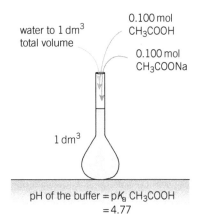

▲ **Figure 1** *Making a buffer*

Question and model answer: Finding the pH of a buffer

A buffer consists of 0.100 mol dm^{-3} ethanoic acid and 0.100 mol dm^{-3} sodium ethanoate (see Figure 1). What is the pH of the buffer? (K_a for ethanoic acid is 1.70×10^{-5}, p$K_a = 4.77$.)

Calculate [H$^+$(aq)] from the equation.

$$K_a = \frac{[H^+(aq)][A^-(aq)]}{[HA(aq)]}$$

Sodium ethanoate is fully dissociated, so [A$^-$(aq)] = 0.100 mol dm^{-3}
Ethanoic acid is almost undissociated, so [HA(aq)] ≈ 0.100 mol dm^{-3}

So $1.70 \times 10^{-5} = [H^+(aq)] \times \frac{0.100}{0.100}$

$1.70 \times 10^{-5} = [H^+(aq)]$

And $pH = -\log_{10}[H^+(aq)]$

$pH = 4.77$

Notice that when you have equal concentrations of acid and salt, the pH of the buffer is equal to the pK_a of the acid used, and this is exactly the same case as the half-neutralisation point.

Summary questions

1. Explain why sulfuric acid and sodium sulfate could not make an acidic buffer. *(1 mark)*

2. Which of the following mixtures could make a basic buffer? **a** ammonia and ammonium bromide, **b** sulfuric acid and potassium sulfate, **c** ammonia and water, **d** sodium hydroxide and sodium chloride. *(1 mark)*

3. What would be the pH of a buffer which is 0.1 mol dm^{-3} ethanoic acid and 0.05 mol dm^{-3} in sodium ethanoate? The pK_a of ethanoic acid is 4.77. *(1 mark)*

Chapter 21 Practice questions

1 Calculate the pH of the following solutions:
   a hydrochloric acid of concentration 0.5 mol dm^{-3} *(1 mark)*
   b sulfuric acid (a diprotic acid) of concentration 0.1 mol dm^{-3} *(1 mark)*
   c benzoic acid (a weak acid, $K_a = 6.3 \times 10^{-5} \text{ mol dm}^{-3}$) of concentration 0.1 mol dm^{-3} *(2 marks)*
   d potassium hydroxide of concentration $0.001 \text{ mol dm}^{-3}$ *(2 marks)*
   e a mixture which is 0.10 mol dm^{-3} in propanoic acid ($K_a = 1.35 \times 10^{-5} \text{ mol dm}^{-3}$) and 0.05 mol dm^{-3} in sodium propanoate. *(2 marks)*

2 a Write an equation for the dissociation of propanoic acid, CH_3CH_2COOH. *(1 mark)*
   b State the expression for the acid dissociation constant, K_a, for propanoic acid. *(1 mark)*
   c The value of K_a for propanoic acid is $1.35 \times 10^{-5} \text{ mol dm}^{-3}$. Calculate the pH of **i** 0.1 mol dm^{-3} solution of propanoic acid. **ii** a buffer solution which is 0.1 mol dm^{-3} in both propanoic acid and sodium propanoate.
   (2 marks each)

3 The pH of a 0.1 mol dm^{-3} solution of ethanoic acid is 2.88. Calculate:
   a $[H^+(aq)]$ in this solution
   b the value of K_a and pK_a for ethanoic acid. *(6 marks)*

22.1 Reactions of Period 3 elements
Specification reference: 3.2.4

The elements of Period 3 are Na, Mg, Al, Si, P, S, Cl, Ar.

The elements – trends
From left to right across the period you go from metals to non-metals. Silicon, in the middle is a semi-metal and conducts electricity somewhat. Argon is a noble gas.

Redox reactions
All the reactions of the elements are redox reactions because the elements always start with oxidation state zero.

> **Revision tip**
> Oxidation states are sometimes called oxidation numbers.

Reactions of the elements
Reaction with water
Sodium reacts vigorously with water, while magnesium reacts slowly with water and vigorously with steam. Make sure you can write balanced equations and work out the oxidation states (shown above the equations).

$$\overset{0}{2Na(s)} + \overset{+1\ -2}{2H_2O(l)} \rightarrow \overset{+1\ -2\ +1}{2NaOH(aq)} + \overset{0}{H_2(g)} \quad \text{pH of resulting solution 13–14}$$

$$\overset{0}{Mg(s)} + \overset{+1\ -2}{2H_2O(l)} \rightarrow \overset{+2\ -2\ +1}{MgOH_2(aq)} + \overset{0}{H_2(g)} \quad \text{the pH of the resulting solution is about 10 – less alkaline because magnesium hydroxide is only sparingly soluble.}$$

> **Revision tip**
> The sum of the oxidation states (taking account of the number of each atom) should be the same on each side of an equation.

Reaction with oxygen
All the elements from sodium to sulfur react directly with oxygen. The metals burn vigorously to form basic oxides while the non-metals form acidic ones. Silicon dioxide is insoluble.

$2Na(s) + \tfrac{1}{2}O_2(g) \rightarrow Na_2O(s)$ – place hot sodium in a gas jar of oxygen.

$Mg(s) + \tfrac{1}{2}O_2(g) \rightarrow MgO(s)$ – place burning magnesium ribbon in a gas jar of oxygen.

$4Al(s) + 3O_2(g) \rightarrow 2Al_2O_3(s)$ – sprinkle aluminium powder into a Bunsen flame.

Everyday aluminium items appear unreactive to air or water. Aluminium is coated with a thin but tough layer of oxide which protects it. Once scratched, the aluminium is so reactive that a new layer of oxide forms immediately.

$Si(s) + O_2(g) \rightarrow SiO_2(s)$

$4P(s) + 5O_2(g) \rightarrow P_4O_{10}(s)$ – some P_2O_3 forms if the supply of oxygen is limited.

$S(s) + O_2(g) \rightarrow SO_2(g)$ – some SO_3 may also form

> **Revision tip**
> Make sure that you can work out the oxidation state of each atom in all these equations.

> **Revision tip**
> Phosphorus has two **allotropes** – red and white. White phosphorus will spontaneously ignite in air.

> **Synoptic link**
> All the metal oxides are white as the metals have no part-filled d-orbitals – see Topic 23.3, Coloured ions.

Summary questions

1. Write a balanced equation for the reaction of magnesium oxide with steam to form magnesium oxide and hydrogen. Include state symbols and oxidation states. *(4 marks)*

2. Show that the sum of the oxidation states of the elements in phosphorus pentoxide is zero. *(2 marks)*

3. Sodium can form another oxide, sodium peroxide, Na_2O_2. State the oxidation state of each atom. *(2 marks)*

22.2 The oxides of elements in Period 3

Specification reference: 3.2.4

Bonding and structure of the oxides

These are summarised in Table 1.

▼ **Table 1** *The trends in the physical properties of some of the oxides of Period 3 elements*

	Na_2O	MgO	Al_2O_3	SiO_2	P_4O_{10}	SO_3	SO_2
T_m / K	1548	3125	2345	1883	573	290	200
Bonding	ionic	ionic	ionic/covalent	covalent	covalent	covalent	covalent
Structure	giant ionic	giant ionic	giant ionic	giant covalent (macromolecular)	molecular	molecular	molecular

As you go from left to right across the period:

- bonding changes from ionic to covalent
- structure changes from giant ionic through macromolecular to small covalent molecules.

The structures are reflected in the melting points – giant structures have high melting points and molecular structures low ones.

The bonding in aluminium oxide has a significant degree of covalent character. This is because the Al^{3+} ion is small and highly charged and can distort the electron cloud around the O^{2-} ion. This is called polarisation, see Figure 1.

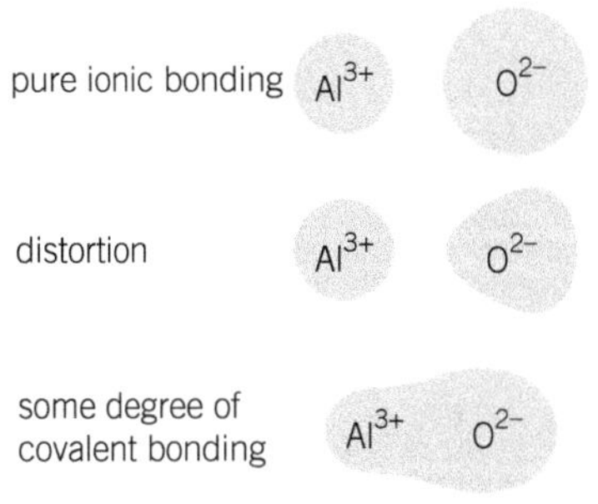

▲ **Figure 1** *Covalency in aluminium oxide*

The reactions of oxides with water

These are summarised in Table 2. There is an overall pattern of alkaline products on the left (metal) side of the Period and acidic products on the right. Aluminium oxide and silicon dioxide are both insoluble in water and do not react.

▼ **Table 2** *The oxides in water*

Oxide	Bonding	Ions present after reaction with water	Acidity/ alkalinity	Approx. pH (Actual values depend on concentration)
Na_2O	ionic	$Na^+(aq), OH^-(aq)$	strongly alkaline	13–14
MgO	ionic	$Mg^{2+}(aq), OH^-(aq)$	somewhat alkaline	10
Al_2O_3	covalent/ ionic	insoluble, no reaction	—	7
SiO_2	covalent	insoluble, no reaction	—	7
P_4O_{10}	covalent	$H^+(aq) + H_2PO_4^-(aq)$	fairly strong acid	1–2
SO_2	covalent	$H^+(aq) + HSO_3^-(aq)$	weak acid	2–3
SO_3	covalent	$H^+(aq) + HSO_4^-(aq)$	strong acid	0–1

The equations are:

$Na_2O(s) + H_2O(l) \rightarrow 2NaOH(aq) \rightarrow 2Na^+(aq) + 2OH^-(aq)$

$MgO(s) + H_2O(l) \rightarrow Mg(OH)_2(aq) \rightleftharpoons Mg^{2+} + 2OH^-(aq)$

$P_4O_{10}(s) + 6H_2O(l) \rightarrow 4H_3PO_4(aq) \rightleftharpoons 4H^+(aq) + 4H_2PO_4^-(aq)$

$SO_2(g) + H_2O(l) \rightarrow H_2SO_3(aq) \rightleftharpoons H^+(aq) + HSO_3^-(aq) \rightleftharpoons 2H^+(aq) + SO_3^{2-}(aq)$

$SO_3(g) + H_2O(l) \rightarrow H_2SO_4(aq) \rightleftharpoons H^+(aq) + HSO_4^-(aq) \rightleftharpoons 2H^+(aq) + SO_4^{2-}(aq)$

When they dissolve in water, the two ionic compounds first produce O^{2-} ions which are strong bases and react with water to form an alkaline solution.

$O^{2-} + H_2O \rightarrow 2OH^-$

The bonding in aluminium oxide ions is too strong for the lattice to dissociate, while silicon dioxide has a giant covalent structure which is not affected by water.

The non-metals produce so-called oxo-acids which dissociate in water to form H^+ ions.

Go further: The chemistry of mortar

Mortar for bricklaying is essentially a mixture of sand and slaked lime, calcium hydroxide, $Ca(OH)_2$, made into a paste with water. The mortar first sets as the water evaporates. Over time, the calcium hydroxide reacts with carbon dioxide in the air to form calcium carbonate (a hard solid) which binds the bricks together.

$Ca(OH)_2(s) + CO_2(g) \rightarrow CaCO_3(s) + H_2O(l)$

a What is the atom economy of this reaction?
b The chemistry of this process is similar to that of a well-known test in school chemistry. Name the test and explain.
Over time, the calcium carbonate in the hardened mortar reacts with more carbon dioxide and water in the air to form soluble calcium hydrogencarbonate, $Ca(HCO_3)_2$, which may be washed away, so the brickwork needs to be re-pointed.
c Write a balanced equation with state symbols for this process.

Summary questions

1 The $H_2PO_4^-$(aq) ion can dissociate further, losing the remaining two hydrogens as H^+ ions. Write the equations for these reactions. *(4 marks)*

2 Give the oxidation states of each atom in the following equation. Is this a redox reaction? Explain.
$Na_2O(s) + H_2O(l) \rightarrow 2NaOH(aq)$ *(3 marks)*

3 Phosphorus forms another oxide, P_4O_6.
 a Suggest how it might be formed from phosphorus. *(1 mark)*
 b Suggest an equation for its reaction with water. *(2 marks)*

22.3 The acidic/basic nature of the Period 3 oxides

Specification reference: 3.2.4

The general trend

The general trend, as in Period 2, is from basic oxides on the left to acidic ones on the right of the Period.

The oxides of sodium and magnesium are basic, while aluminium oxide shows both acidic and basic properties – it is called an amphoteric oxide. Silicon, phosphorus, and sulfur oxides get increasingly acidic. Chlorine forms several oxides which are not considered here – however they are acidic as expected. Argon does not form an oxide.

The basic oxides

The oxides of sodium and magnesium both react with acids to form a salt and water, for example

$Na_2O(s) + 2HCl(aq) \rightarrow 2NaCl(aq) + H_2O(l)$

$MgO(s) + H_2SO_4(aq) \rightarrow MgSO_4(aq) + H_2O(aq)$

The amphoteric oxide

Aluminium is a metal but it shows some non-metal characteristics.

Aluminium oxide will react with both acids and bases: with hydrochloric acid

$Al_2O_3(s) + 6HCl(aq) \rightarrow 2AlCl_3(aq) + 3H_2O(l)$

and with hot concentrated sodium hydroxide

$Al_2O_3(s) + 2NaOH(l) + 3H_2O \rightarrow 2NaAl(OH)_4(aq)$

The product is sodium aluminate.

The acidic oxides

Silicon dioxide is weakly acidic – it will react with hot, concentrated strong alkalis for example

$SiO_2(s) + 2NaOH(aq) \rightarrow Na_2SiO_3(aq) + 3H_2O(l)$

The product is sodium silicate.

Phosphorus pentoxide first reacts with water to form phosphoric(V) acid, H_3PO_4, which then reacts in stages with an alkali:

The overall reaction is:

$3KOH(aq) + H_3PO_4(aq) \rightarrow K_3PO_4 + H_2O(l)$

Sulfur dioxide reacts with sodium hydroxide to form first sodium hydrogensulfate(IV) and then sodium sulfate(IV):

$SO_2(g) + NaOH(aq) \rightarrow NaHSO_3(aq)$

$NaHSO_3(aq) + NaOH(aq) \rightarrow Na_2SO_3(aq) + H_2O(l)$

> **Revision tip**
> An alkali is a soluble base.

> **Revision tip**
> Make sure that you can write balanced equations for each oxide reacting with any common acid.

> **Revision tip**
> Sodium hydrogensulfate(IV) and sodium sulfate(IV) are often called sodium hydrogensulfite and sodium sulfite respectively.

Summary questions

1. The gas sulfur trioxide, SO_3, also reacts with alkalis. Write a balanced symbol equation with state symbols for its overall reaction with potassium hydroxide. *(2 marks)*

2. Give the oxidation states of each atom in the equation
$3KOH(aq) + H_3PO_4(aq) \rightarrow K_3PO_4 + 3H_2O(l)$ *(4 marks)*

3. Write balanced symbol equations with state symbols for the reactions of sodium oxide with sulfuric acid and magnesium oxide with hydrochloric acid. *(4 marks)*

Chapter 22 Practice questions

1. **a** Write equations for the reactions that occur when **i** sodium oxide and **ii** sulfur dioxide react with water. Show the ions that are formed.
   (4 marks)

   b State the trend in the properties of the oxides that this illustrates as you go across Period 3 in the Periodic Table. *(1 mark)*

   c Aluminium oxide is described as amphoteric. Explain this term and give equations that illustrate this behaviour. *(5 marks)*

2. An industrial process produces waste containing phosphorus(V) oxide (phosphorus pentoxide, P_4O_{10}). This is reacted with water to form an acidic solution which is neutralised before further disposal.

   a Write an equation for the reaction of phosphorus pentoxide with water.
   (2 marks)

   b The resulting acidic solution is neutralised with excess magnesium oxide.

   i What is meant by 'excess'? *(1 mark)*

   ii Write an equation for this neutralisation. *(2 marks)*

   iii State the oxidation state of each atom in the equation before and after reaction. Is this a redox reaction? *(4 marks)*

   c What happens to the excess magnesium oxide and how could it be separated from the neutralised waste water? *(2 marks)*

   d Why is this scheme better than using excess sodium hydroxide to neutralise the acid? *(1 mark)*

3. Consider the following oxides of elements in Period 3: Na_2O, MgO, Al_2O_3, SiO_2, P_4O_{10}, SO_3.

   a Select one that is amphoteric. *(1 mark)*

   b Select two that form acidic solutions in water. *(2 marks)*

   c Select one with a giant covalent structure. *(1 mark)*

   d Select one that is a gas. *(1 mark)*

   e Write an equation for the reaction that happens when **i** Na_2O and **ii** SO_3 react with water. *(4 marks)*

   f Write an equation for the reaction that happens when the products of **e i** and **ii** react together. *(2 marks)*

23.1 The general properties of transition metals

Specification reference: 3.2.5

The elements from titanium to copper are called transition metals. They are included in the d-block elements and outlined in Figure 1.

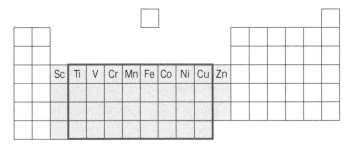

▲ **Figure 1** *The d-block elements (shaded) and the transition metals (outlined)*

The transition elements are all typical metals. Their chemical properties are very similar and this can be explained by their electronic configuration.

What is a transition element?

A transition element forms at least one stable ion with a *part-full* d-shell of electrons.

Not all d-block elements are transition metals. Scandium only forms Sc^{3+} ($3d^0$), and zinc only forms Zn^{2+} ($3d^{10}$) in all their compounds. They are therefore d-block elements but not transition elements.

Figure 2 shows the electron arrangements for the elements in the first row of the d-block.

In general there are two outer 4s electrons and as you go across the period, electrons are added to the *inner* 3d sub-shell. This explains the overall similarity of these elements.

Chromium, Cr, and copper, Cu, have only one electron in the 4s outer shell. In copper, Cu, the d sub-shell is full ($3d^{10}$) and in chromium, Cr, it is half full ($3d^5$), which makes them more stable than expected, but they both can form ions with a part full d-shell.

Chemical properties of transition metals

The chemistry of transition metals has four main features which are common to all the elements:

- Complex formation: A complex ion is formed when a transition metal ion is surrounded by ions or other molecules, collectively called ligands. See Topic 23.2, Complex formation and the shape of complex ions.
- Colour: Most transition metal ions are coloured. See Topic 23.3, Coloured ions.
- Variable oxidation states: They have more than one oxidation state in their compounds. They can therefore take part in many redox reactions. See Topic 23.4, Variable oxidation states of transition elements.
- Catalysis: Many transition metals, and their compounds, show catalytic activity. See 23.5, Catalysts.

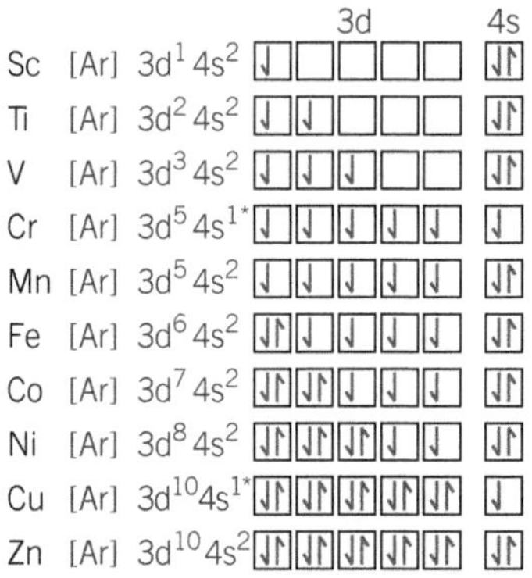

			3d	4s
Sc	[Ar]	$3d^1 4s^2$	↓□□□□	↓↑
Ti	[Ar]	$3d^2 4s^2$	↓↓□□□	↓↑
V	[Ar]	$3d^3 4s^2$	↓↓↓□□	↓↑
Cr	[Ar]	$3d^5 4s^1$*	↓↓↓↓↓	↓
Mn	[Ar]	$3d^5 4s^2$	↓↓↓↓↓	↓↑
Fe	[Ar]	$3d^6 4s^2$	↓↑↓↓↓↓	↓↑
Co	[Ar]	$3d^7 4s^2$	↓↑↓↑↓↓↓	↓↑
Ni	[Ar]	$3d^8 4s^2$	↓↑↓↑↓↑↓↓	↓↑
Cu	[Ar]	$3d^{10}4s^1$*	↓↑↓↑↓↑↓↑↓↑	↓
Zn	[Ar]	$3d^{10}4s^2$	↓↑↓↑↓↑↓↑↓↑	↓↑

▲ **Figure 2** *The electronic arrangement of the elements in the first transition series. [Ar] represents $1s^2\ 2s^2\ 2p^6\ 3s^2\ 3p^6$*

Revision tip
When transition metals react to form ions, the 4s electrons are always lost first.

Summary questions

1. Why does the fact that the d-shell is filling after the s-shell mean the transition metals are similar? *(1 mark)*

2. **a** Write the full electron configuration for a titanium atom in the form $1s^2$ etc. *(1 mark)*
   b Write the electron configuration of a titanium atom using the shorthand [Ar] etc. *(1 mark)*
   c Write the electron configuration of a Ti^{2+} ion using the shorthand [Ar] etc. *(1 mark)*

3. The electron configuration of the manganese atom is [Ar] $3d^5 4s^2$.
   a Which electrons are removed when a Mn^{2+} ion is formed? *(1 mark)*
   b Suggest why this ion would be expected to be particularly stable. *(1 mark)*

23.2 Complex formation and the shape of complex ions

Specification reference: 3.2.5

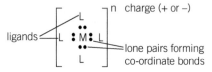

▲ **Figure 1** *A complex ion of a metal M with co-ordination number 4*

The formation of complex ions

All transition metal ions can accept electrons from ligands to form co-ordinate bonds.

Some examples of ligands are $H_2O:$, $:NH_3$, $:Cl^-$, $:CN^-$

Two, four, or six ligands can bond to a transition metal ion to form a **complex ion**. The number of ligands that surround the metal ion is called the **co-ordination number**.

Multidentate ligands – chelation

Multidentate ligands have more than one atom with a lone pair of electrons which can bond to a transition metal ion. Bidentate ligands have two lone pairs, see Figures 2 and 3.

The ion ethylenediaminetetraacetate, called $EDTA^{4-}$, see Figure 4, can act as a hexadentate ligand.

The resulting complex ions with their multidentate ligands are called **chelates**.

The chelate effect

If you add a hexadentate ligand such as EDTA to a solution of a transition metal salt, the EDTA will replace all six water ligands in an aqua ion such as $[Cu(H_2O)_6]^{2+}$:

$[Cu(H_2O)_6]^{2+}(aq) + EDTA^{4-}(aq) \rightarrow [CuEDTA]^{2-}(aq) + 6H_2O(l)$

The entropy of the system increases from left to right, which is why $[CuEDTA]^{2-}(aq)$ is readily formed.

Haemoglobin

Haemoglobin is a compound of Fe^{2+} complexed with a tetradentate ring ligand, haem, and a protein, globin. A sixth ligand site accepts an oxygen molecule, which is a poor ligand. The complex carries oxygen round the bloodstream, releasing it in the cells. But oxygen can be replaced by carbon monoxide, which is a better ligand, and this prevents the carriage of oxygen.

Shapes and charges of complex ions

- Ions with co-ordination number six are octahedral, for example, $[Co(NH_3)_6]^{3+}$.
- Ions with co-ordination number four are usually tetrahedral, for example, $[CoCl_4]^{2-}$.
- Some ions with co-ordination number four are square planar, for example, $[Ni(CN)_4]^{2-}$.
- Complex ions may have a positive charge or a negative charge.

Key terms

Co-ordinate (dative) bond: A covalent bond formed when the pair of electrons comes from the same atom.

Ligand: An ion or molecule that donates a lone pair of electrons to form a co-ordinate bond with a transition metal.

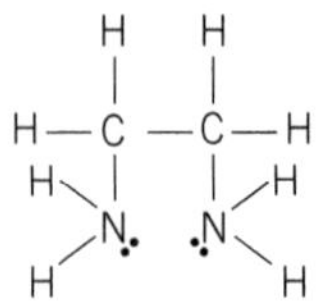

▲ **Figure 2** *Ethane 1,2-diamine (abbreviation en)*

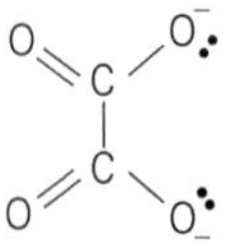

▲ **Figure 3** *The ethanedioate ion*

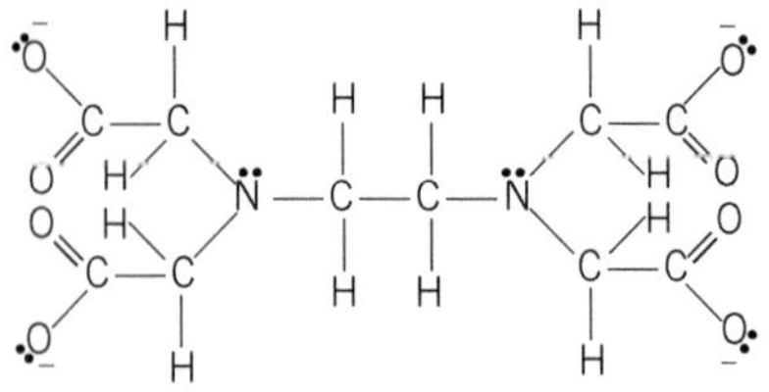

▲ **Figure 4** *EDTA*

Revision tip

When the ligand is water, the complex ion is called an aqua ion, for example, $[Cu(H_2O)_6]^{2+}(aq)$.

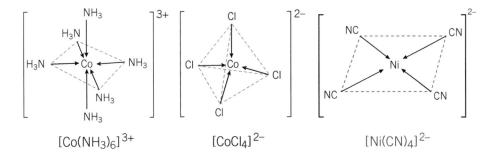

In $[Co(NH_3)_6]^{3+}$, the metal ion has a charge of +3 and as the NH_3 ligands are all neutral, the complex ion has an overall charge of +3.

In $[CoCl_4]^{2-}$, the metal ion has a charge of +2 and each of the four ligands Cl^- has a charge of −1, so the complex ion has an overall charge of −2.

The reason for the difference in shape is that the chloride ion is a larger ligand than the ammonia molecule, so only four Cl^- ligands can fit around the metal ion, whereas there is room for six NH_3 ligands.

Some complexes are linear – one example is $[Ag(NH_3)_2]^+$:

$[H_3N \rightarrow Ag \leftarrow NH_3]^+$

A solution containing this complex ion is called Tollens' reagent and is used in organic chemistry to distinguish aldehydes from ketones, see Topic 26.2, Reactions of the carbonyl group in aldehydes and ketones.

 Go further: The shape of xenon tetrafluoride

Xenon tetrafluoride, XeF_4, is one of a small number of inert gas compounds. The factors governing their shapes are similar to those for transition metal complexes. The dot-and-cross diagram for XeF_4 is shown.

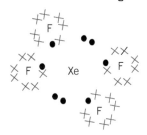

a In XeF_4, the xenon atom has more than eight electrons in its full outer shell. Explain how this can occur.
b How many electrons does the xenon atom have in its outer shell in XeF_4?
c What shape is the XeF_4 molecule based on?
d How many lone pairs are there in the outer shell of the xenon atom in XeF_4?

The structure of XeF_4 is shown.

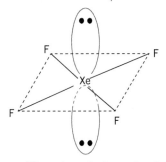

e What shape is the molecule?
f Explain why the lone pairs take up the positions that they do.

The transition metals

> **Revision tip**
> Isomers are compounds with the same molecular formula but with different arrangements of their atoms in space.

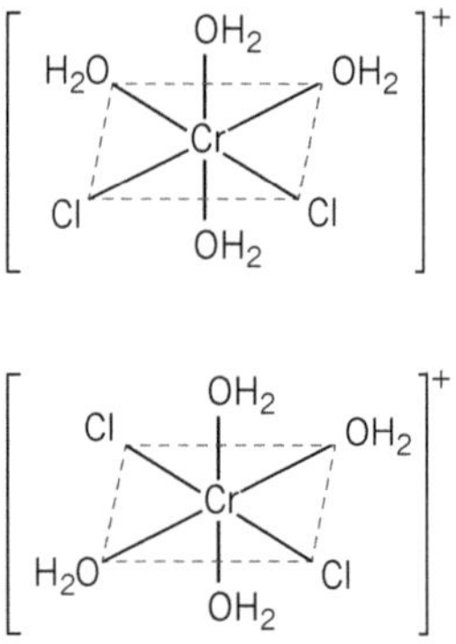

[Structural diagrams of cis- and trans- isomers of [CrCl₂(H₂O)₄]⁺]

▲ **Figure 5** *The cis- or Z- isomer (top) and the trans- or E- isomer (bottom)*

> **Revision tip**
> Optical isomers are non-superimposable mirror images of each other.

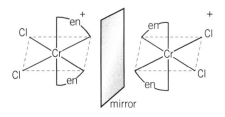

en is an abbreviation for ethane-1,2-diamine:

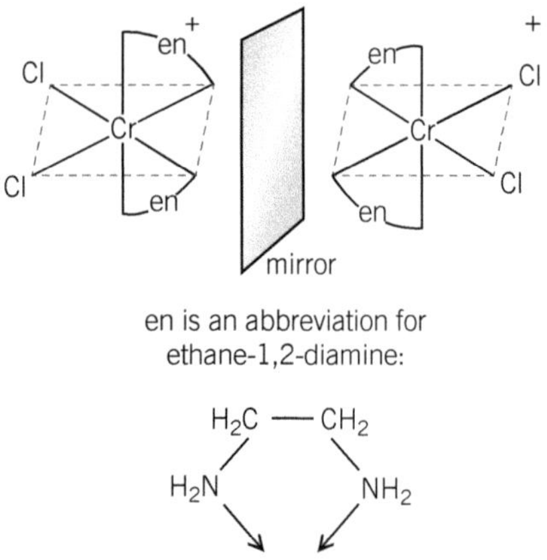

▲ **Figure 6** *Transition metal complexes that are non-identical mirror images of each other (top) and the structure of the ethane-1,2,-diamine (en) ligand (bottom)*

Isomerism in transition metal complexes

Transition metal complexes can form both geometrical isomers (*cis-trans* or *E-Z* isomers) and optical isomers.

Geometrical isomerism

This occurs in octahedral and square planar complexes. For example in the complex ion $[CrCl_2(H_2O)_4]^+$, the Cl⁻ ligands may be next to each other (the *cis-* or *Z-* form) or on opposite sides of the central chromium ion (the *trans-* or *E-*form) (Figure 5).

In the square planar complex platin, the Cl⁻ ligands may be next to each other (the *cis-* or *Z-* form) or on opposite sides of the central platinum atom (the *trans-* or *E-* form).

Optical isomerism

Transition metal complexes exist as pairs of optical isomers if they have two or more bidentate ligands (see Figure 6).

Optical isomers are chiral. They have identical chemical properties but one isomer will rotate the plane of polarisation of polarised light clockwise and the other anticlockwise.

> **Summary questions**
>
> 1 $[Cu(H_2O)_6]^{2+}(aq) + EDTA^{4-}(aq) \rightarrow [CuEDTA]^{2-}(aq) + 6H_2O(l)$
>   a How many entities are there on each side of the equation? *(1 mark)*
>   b Explain why this causes an increase in entropy. *(1 mark)*
>
> 2 Explain why carbon monoxide is a poison, even when there may be oxygen present in the air. *(2 marks)*
>
> 3 a Which is the larger ligand, NH_3 or Br^-? *(1 mark)*
>   b How does this affect the shape of a complex ion? *(2 marks)*
>
> 4 This is the structure of a ligand. 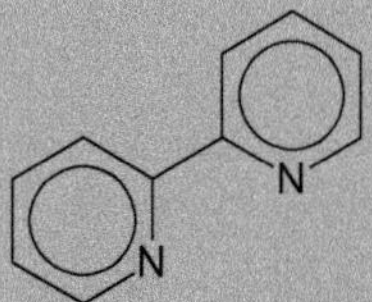
>
>   a Suggest which atoms are most likely to form co-ordinate bonds. *(1 mark)*
>   b What feature of these atoms allows them to form co-ordinate bonds? *(1 mark)*
>   c Is it likely to be a monodentate, bidentate, or hexadentate ligand? *(1 mark)*

23.3 Coloured ions

Specification reference: 3.2.5

Most transition metal compounds are coloured.

If a solution of a substance looks purple, it is because it absorbs all the light from a beam of white light except red and blue. The red and blue light passes through and the solution appears purple, (Figure 1).

Why are transition metal complexes coloured?

- Transition metal compounds have part-filled d-orbitals so electrons are able to move from one d-orbital to another.
- The presence of other atoms nearby means the d-orbitals have slightly different energies.
- Electrons move from one d-orbital to another of a higher energy level (an excited state). As they do so, they absorb energy equal to the difference in energy, ΔE, between levels. This is often in the visible region of the spectrum.
- This colour is therefore missing from the spectrum and you see the combination of the colours that are not absorbed.

ΔE is responsible for the colour of the complex. It depends on the oxidation state of the metal and also on the ligands (and therefore the shape of the complex ion).

$$\Delta E = h\nu = \frac{hc}{\lambda}$$ where ν is frequency, λ is wavelength, c is the velocity of light, and h is Planck's constant.

Table 1 shows examples of how the oxidation state of the metal affects the colour of the complex.

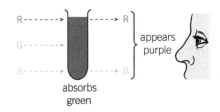

▲ **Figure 1** *Solutions look coloured because they absorb some colours and let others pass through*

▼ **Table 1**

Oxidation state of metal	2	3
iron complexes	$[Fe(H_2O)_6]^{2+}$ green	$[Fe(H_2O)_6]^{3+}$ pale brown
chromium complexes	$[Cr(H_2O)_6]^{2+}$ blue	$[Cr(H_2O)_6]^{3+}$ red-violet
cobalt complexes	$[Co(NH_3)_6]^{2+}$ brown	$[Co(NH_3)_6]^{3+}$ yellow

Colorimetry

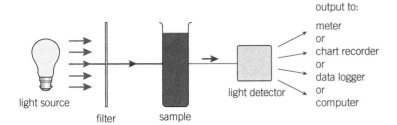

▲ **Figure 2** *A simple colorimeter*

You can use a colorimeter, see Figure 2, to measure the concentration of coloured solutions. The more concentrated the colour, the less light will reach the detector. A suitable filter makes the experiment more sensitive.

Revision tip
The absorption of visible light is used in spectroscopy.

Summary questions

1. Explain why solutions containing the Sc^{3+} ion are colourless, although scandium is in the d-block. *(1 mark)*

2. Solutions containing V^{3+} ions are green.
   a What colour of light passes through this solution? *(1 mark)*
   b What happens to the other colours of the visible spectrum? *(1 mark)*

3. $[CuCl_4]^{2-}$ is yellow/green whereas $[Cu(NH_3)_6]^{2+}$ is dark blue. Suggest two factors that could be responsible for the difference in colour. *(2 marks)*

23.4 Variable oxidation states of transition elements

Specification reference: 3.2.5

▼ **Table 1** *Zinc ions in acid solution will reduce vanadium(V) through oxidation states V(IV) and V(III) to V(II).*

Oxidation number	Species	Colour
5	$VO_2^+(aq)$	yellow
4	$VO^{2+}(aq)$	blue
3	$V^{3+}(aq)$	green
2	$V^{2+}(aq)$	violet

Revision tip
Remember the phrase OIL RIG — oxidation is loss, reduction is gain (of electrons).

Oxidation states

A typical transition metal can use its 3d electrons as well as its 4s electrons in bonding, and so can have a variety of oxidation states in different compounds. All the transition elements show the +2 oxidation state. These are ions formed by the loss of 4s electrons.

The higher oxidation states are not simple ions. For example, Mn^{7+} ions only exist covalently bonded to oxygen in a compound ion as in MnO_4^-.

Table 1 shows the colour changes when the vanadate(V) ion is reduced.

Redox titrations

The concentration of an oxidising or a reducing agent can be measured with a redox titration provided we know the equation for the reaction.

Question and model answer: What is the equation for the reaction of MnO_4^- with Fe^{2+}?

During this reaction the oxidation number of the manganese falls from +7 to +2.

1. $MnO_4^-(aq) \rightarrow Mn^{2+}(aq)$ — *Write the half equation for the reduction of Mn(VII) to Mn(II).*

2. $MnO_4^- + 8H^+(aq) \rightarrow Mn^{2+}(aq) + 4H_2O(l)$ — *Balance the oxygen atoms using H^+ ions and H_2O molecules.*

3. $MnO_4^-(aq) + 5e^- + 8H^+(aq) \rightarrow Mn^{2+}(aq) + 4H_2O(l)$ — *Balance for charge using electrons.*

4. $Fe^{2+}(aq) \rightarrow Fe^{3+}(aq) + e^-$ — *Write the half equation for the oxidation of iron(II) to iron(III).*

5. $5Fe^{2+}(aq) \rightarrow 5Fe^{3+}(aq) + 5e^-$

$MnO_4^-(aq) + 5e^- + 8H^+(aq) \rightarrow Mn^{2+}(aq) + 4H_2O(l)$

Multiply the Fe^{2+}/Fe^{3+} half reaction by five (so that the numbers of electrons in each half reaction are the same) and then add the two half equations.

$5Fe^{2+}(aq) + MnO_4^-(aq) + \cancel{5e^-} + 8H^+(aq) \rightarrow 5Fe^{3+}(aq) + \cancel{5e^-} + Mn^{2+}(aq) + 4H_2O(l)$

$5Fe^{2+}(aq) + MnO_4^-(aq) + 8H^+(aq) \rightarrow 5Fe^{3+}(aq) + Mn^{2+}(aq) + 4H_2O(l)$

Cancel out the electrons on each side

So, $Fe^{2+}(aq)$ reacts with manganate(VII) ions in the ratio 5 : 1.

The colour of the mixture changes from almost colourless Fe^{2+} to pale pink Mn^{2+} as the reaction proceeds. At the end point, one more drop of $MnO_4^-(aq)$ ions will turn the solution purple. So no indicator is needed for this titration.

The transition metals

Worked example: The reaction between manganate(VII) ions and ethanedioate ions

To find the redox equation use the steps as before

1, 2, and 3 $MnO_4^-(aq) + 8H^+(aq) + 5e^- \rightarrow Mn^{2+}(aq) + 4H_2O(l)$

4 $C_2O_4^{2-}(aq) \rightarrow 2CO_2(g) + 2e^-$

Multiply first equation by 2 and second by 5 to match the number of electrons and then add the two equations together

$2MnO_4^-(aq) + 16H^+(aq) + 5C_2O_4^{2-}(aq) \rightarrow 10CO_2(g) + 2Mn^{2+}(aq) + 8H_2O(l)$

Summary questions

1 In what ratio do $MnO_4^-(aq)$ ions react with $C_2O_4^{2-}(aq)$ ions? *(1 mark)*

2 **a** Add the oxidation states to this equation.
$2MnO_4^-(aq) + 16H^+(aq) + 5C_2O_4^{2-}(aq) \rightarrow 10CO_2(g) + 2Mn^{2+}(aq) + 8H_2O(l)$
(1 mark)
b Which atoms have been oxidised and which reduced? *(1 mark)*

3 Use the half equation technique to write a balanced equation for the reduction of VO_2^+ to VO^{2+} by zinc in acid solution.
$(Zn(s) \rightarrow Zn^{2+}(aq) + 2e^-)$. *(1 mark)*

23.5 Catalysis
Specification reference: 3.2.5

> **Key term**
>
> **Catalyst:** Catalysts speed up reactions without being permanently chemically changed or used up.

> **Synoptic link**
>
> Catalysts and activation energy are introduced in Topic 5.3, Catalysts.

> **Revision tip**
>
> Adsorption means forming weak bonds with a strength somewhere between covalent bonds and van der Waals forces.

> **Summary questions**
>
> 1 What steps do you have to go through to deduce the overall equation for the oxidation of sulfur dioxide from the two steps shown above? *(2 marks)*
>
> 2 What happens to the oxidation state of the vanadium pentoxide as it catalyses the oxidation of sulfur dioxide? *(2 marks)*
>
> 3 a Add the two steps of the oxidation of iodide ions to iodine by peroxodisulfate ions to find the overall reaction. *(2 marks)*
>   b Select from 'oxidising agent', 'acid', 'reducing agent',' base'. Which term best describes the function of **i** Fe^{2+}, **ii** Fe^{3+}? *(2 marks)*
>   c Suggest why the two steps are faster than the uncatalysed reaction. *(2 marks)*

Catalysts lower the activation energy for reactions.

Homogeneous and heterogeneous catalysts

- **Homogeneous catalysts** are in the same phase as the reactants – usually in aqueous solutions.
- **Heterogeneous catalysts** are in a different phase to the reactants – usually a solid catalyst for a gas phase reaction.

Many metal catalysts work at the surface by forming weak bonds with the molecules in a reaction. This is called adsorption. It weakens the bonds in the reactants allowing the reaction to proceed more easily. The metal also holds the reactants in the correct orientation to react. Transition metals use their part-filled d-orbitals to form weak bonds with the reactants. These bonds must be strong enough to hold the reactants but weak enough to release the products.

Catalysts may also work by using the property of a transition metal to have variable oxidation states. For example, vanadium(V) oxide (vanadium pentoxide, V_2O_5) catalyses the oxidation of sulfur dioxide, SO_2, to sulfur trioxide, SO_3, by oxygen in the Contact process for making sulfuric acid. The steps are:

$$SO_2(g) + V_2O_5(s) \rightarrow SO_3(g) + V_2O_4(s)$$

$$2V_2O_4(s) + O_2(g) \rightarrow 2V_2O_5$$

The overall reaction is $2SO_2(g) + O_2(g) \rightarrow 2SO_3(g)$

Examples of heterogeneous catalysts

Iron in the Haber process for making ammonia.

Platinum and rhodium in catalytic converters for car exhaust systems. These convert polluting gases such as carbon monoxides, nitrogen oxides, and hydrocarbons into nitrogen, carbon dioxide, and water.

Examples of homogeneous catalysis

The oxidation of iodide ions to iodine by peroxodisulfate ions, $S_2O_8^{2-}$, catalysed by Fe^{2+} ions.

Step 1: $S_2O_8^{2-}(aq) + 2Fe^{2+}(aq) \rightarrow 2SO_4^{2-}(aq) + 2Fe^{3+}(aq)$

Step 2: $2Fe^{3+}(aq) + 2I^- \rightarrow 2Fe^{2+} + I_2(aq)$

The catalyst is regenerated in step 2.

Autocatalysis

This occurs when a product of the reaction acts as catalyst. The reaction starts slowly and then speeds up once the catalyst is formed. For example, in the following reaction, Mn^{2+} is the catalyst. Once the reaction starts the Mn^{2+} reacts with MnO_4^- ions to form Mn^{3+} as an intermediate species. This then reacts with $C_2O_4^{2-}$ ions to reform Mn^{2+}.

$$2MnO_4^-(aq) + 16H^+(aq) + 5C_2O_4^{2-}(aq) \rightarrow 10CO_2(g) + 2Mn^{2+}(aq) + 8H_2O(l)$$

Chapter 23 Practice questions

1. **a** Complete the electron arrangement of **i** the Cr^{3+} ion **ii** the Zn^{2+} ion. *(2 marks)*

   $1s^2$ ......

   b Explain why chromium is considered to be a transition metal and zinc is not. *(1 mark)*

   c The complex ion $[CrCl_2(H_2O)_4]^+$ exists as two isomers. One has the structure

   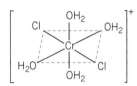

   i Draw the structure of the other isomer in the same style. *(1 mark)*

   ii What type of isomerism is this? *(1 mark)*

   iii Which isomer is the one drawn above? *(1 mark)*

2. **a** State the shape of the $[Cu(H_2O)_6]^+$ ion and the $CuCl_4^{2-}$ ion. Explain why they have different shapes and different charges. *(6 marks)*

   b Predict the shape of the $CuBr_4^{2-}$ ion and explain your reasoning. *(2 marks)*

3. **a** Complete the electron arrangement of the Cu^{2+} ion starting from [Ar]. *(1 mark)*

   b Explain in terms of the d orbitals why the $[Cu(H_2O)_6]^{2+}$ ion is coloured. *(4 marks)*

   c i State the meaning of each term in the equation $\Delta E = hv$. **ii** Explain the relevance of this equation to the colour of a transition metal ion. *(4 marks)*

4.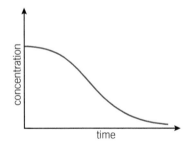

   The diagram shows a reaction rate curve for a reaction involving transition metal ions.

   a What is unusual about the shape of the curve? *(2 marks)*

   b Suggest what type of reaction is occurring and describe how this explains the shape of the curve. *(3 marks)*

24.1 The acid–base chemistry of aqueous transition metal ions

Specification reference: 3.2.6

> **Revision tip**
> When drawing co-ordinate bonds, the arrow represents the donated pair of electrons.

Transition metals form complexes with ligands. When the ligand is water, metal complexes are called **aqua ions**. For example, $[Cu(H_2O)_6]^{2+}$ is present in copper sulfate solutions, see Topic 23.2, Complex formation and the shape of complex ions.

In a solution of iron(II) nitrate, $Fe(NO_3)_2$, the Fe^{2+} ion exists as the complex ion $[Fe(H_2O)_6]^{2+}$. In a solution of an iron(III) salt the complex formed is $[Fe(H_2O)_6]^{3+}$. These are both octahedral.

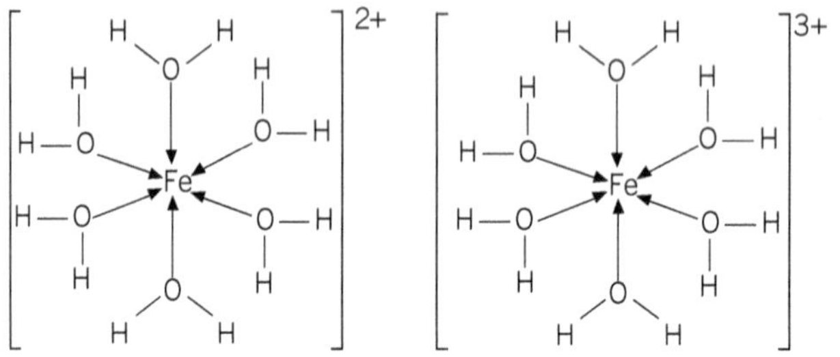

▲ **Figure 1** $[Fe(H_2O)_6]^{2+}$ (left) and $[Fe(H_2O)_6]^{3+}$ (right)

The acidity of metal aqua ions in aqueous solution

> **Revision tip**
> The *smaller* the value of pK_a, the *stronger* the acid.

With transition metals, M^{3+} (aq) ions are more acidic than M^{2+} (aq) ions. For example, Fe^{2+}(aq) solutions are not noticeably acidic, but those of Fe^{3+}(aq) ($pK_a = 2.2$) are stronger acids than ethanoic acid ($pK_a = 4.8$).

The Fe^{3+} ion is both smaller and more highly charged than Fe^{2+} and strongly attracts electrons from the oxygen atoms of the water ligands and this weakens the O—H bonds. The complex ion will then release a H^+ ion making the solution acidic (Figure 2). Fe^{2+} is less polarising and so fewer O—H bonds break in solution.

$[Fe(H_2O)_6]^{3+}(aq) \rightleftharpoons [Fe(H_2O)_5(OH)]^{2+}(aq) + H^+(aq)$

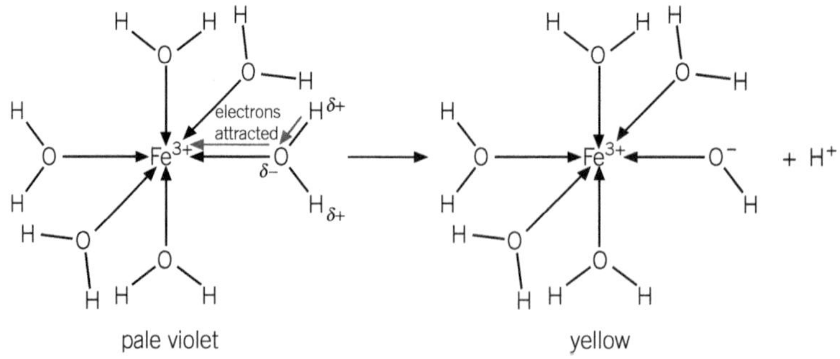

▲ **Figure 2** *The acidity of* Fe^{3+}(aq) *ions*

Amphoteric hydroxides

Amphoteric means showing both acidic and basic properties.

Aluminium hydroxide, $Al(H_2O)_3(OH)_3$, is amphoteric. For example, with hydrochloric acid it acts as a base:

$Al(H_2O)_3(OH)_3 + 3HCl \rightarrow Al(H_2O)_6^{3+} + 3Cl^-$

But aluminium hydroxide, like Fe^{3+}, has a small and highly charged metal ion and will react with the base sodium hydroxide to give a colourless solution of sodium tetrahydroxoaluminate:

$Al(H_2O)_3(OH)_3 + OH^- \rightarrow [Al(OH)_4]^- + 3H_2O$

That is, is acting as an acid.

> **Summary questions**
>
> 1. Why are transition metals M^{3+} (aq) ions more acidic that M^{2+} (aq) ions?
>   *(1 mark)*
>
> 2. a What would you expect to see if sodium carbonate is added to a solution containing iron(III) aqua ions?
>   *(1 mark)*
>   b Explain your answer.
>   *(1 mark)*
>
> 3. Explain why the *smaller* the value of pKa, the *stronger* the acid.
>   *(2 marks)*

24.2 Ligand substitution reactions
24.3 A summary of acid–base and substitution reactions of some metal ions

Specification reference: 3.2.6

Some test tube reactions of metal ions

The products of many of the reactions of transition metal compounds can be identified by their colours. Table 1 and Table 2 show a number of examples of reactions of metal ions with different bases. The effect of adding the base shown on the left to solutions of aqua ions is shown in the following tables.

▼ **Table 1** *Adding bases to M^{2+}(aq) complexes*

	$[Fe(H_2O)_6]^{2+}$(aq) pale green	$[Cu(H_2O)_6]^{2+}$(aq) pale blue
OH^- little	green gelatinous ppt of $[Fe(H_2O)_4(OH)_2]$*	pale blue ppt of $[Cu(H_2O)_4(OH)_2]$
OH^- excess	green gelatinous ppt of $[Fe(H_2O)_4(OH)_2]$*	pale blue ppt of $[Cu(H_2O)_4(OH)_2]$
NH_3 little	green gelatinous ppt of $[Fe(H_2O)_4(OH)_2]$*	pale blue ppt of $[Cu(H_2O)_4(OH)_2]$
NH_3 excess	green gelatinous ppt of $[Fe(H_2O)_4(OH)_2]$*	deep blue solution of $[Cu(NH_3)_4(H_2O)_2]^{2+}$
CO_3^{2-}	green ppt of $FeCO_3$	blue-green ppt of $CuCO_3$

*Pale green $[Fe(H_2O)_4(OH)_2]$ is soon oxidised by air to brown $[Fe(H_2O)_3(OH)_3]$

In most cases the solid metal hydroxide is precipitated $[M(H_2O)_4(OH)_2]$. Ammonia is also a base *and* a good ligand and, in the case of copper, the copper hydroxide dissolves in excess ammonia forming the deep blue ion $[Cu(NH_3)_4(H_2O)_2]^{2+}$ by a ligand substitution reaction. Precipitates of the metal carbonates form when carbonate ions are added.

▼ **Table 2** *Adding bases to M^{3+}(aq) complexes*

	$[Fe(H_2O)_6]^{3+}$(aq) purple/yellow/brown	$[Al(H_2O)_6]^{3+}$(aq) colourless
OH^- little	brown gelatinous ppt of $[Fe(H_2O)_3(OH)_3]$	white ppt of $[Al(H_2O)_3(OH)_3]$
OH^- excess	brown gelatinous ppt of $[Fe(H_2O)_3(OH)_3]$	colourless solution of $[Al(OH)_4]^-$
NH_3 little	brown gelatinous ppt of $[Fe(H_2O)_3(OH)_3]$	white ppt of $[Al(H_2O)_3(OH)_3]$
NH_3 excess	brown gelatinous ppt of $[Fe(H_2O)_3(OH)_3]$	white ppt of $[Al(H_2O)_3(OH)_3]$
CO_3^{2-}	brown gelatinous ppt of $[Fe(H_2O)_3(OH)_3]$ and bubbles of CO_2	white ppt of $[Al(H_2O)_3(OH)_3]$ and bubbles of CO_2

In the case of the M^{3+} ions, bubbles of carbon dioxide are produced rather than the solid carbonate. This is because $[M(H_2O)_6]^{3+}$ is more acidic than $[M(H_2O)_6]^{2+}$, see Topic 24.1, The acid–base chemistry of aqueous transition metal ions.

Summary questions

1. Write an equation for copper hydroxide dissolving in ammonia solution. *(2 marks)*

2. Explain why solid $FeCO_3$ exists, but solid $Fe_2(CO_3)_3$ does not. *(1 mark)*

3. Give the electron arrangement of Al^{3+} in the form $1s^2$ etc. and use it to explain why aluminium compounds are colourless. *(2 marks)*

Chapter 24 Practice questions

1. Adding a solution of sodium carbonate to a solution containing aqueous Fe^{2+} ions produces a green precipitate. Adding a solution of sodium carbonate to a solution containing aqueous Fe^{3+} ions produces a brown precipitate and bubbles of a gas.

   a. Identify the products *(2 marks)*

   b. Explain the difference between the behaviour of Fe^{3+} and Fe^{2+} ions. *(2 marks)*

   c. Give appropriate equations. *(6 marks)*

2. a. Explain the term **chelate**. *(1 mark)*

   b. The ligand $EDTA^{4-}$ is a hexadentate ligand. Write an equation to show how $EDTA^{4-}$ can replace the H_2O ligands in the ion $[Cu(H_2O)_6]^{2+}$. *(2 marks)*

   c. How many entities are there on **i** the left and **ii** the right of the equation? *(2 marks)*

   d. Explain why this means the reaction goes to completion. *(2 marks)*

3. Consider the following reactions.

   $$[Cu(H_2O)_6]^{2+}(aq) \xrightarrow{\text{Conc. HCl}} A \quad (1)$$

   $[Cu(H_2O)_6]^{2+} \rightarrow$ blue-green precipitate B   (2)

   $[Cu(H_2O)_6]^{2+} \rightarrow$ pale blue precipitate C $\rightarrow [Cu(NH_3)_6]^{2+}(aq)$ (3)

   a. Give the formula of the complex ion A. *(1 mark)*

   b. **i** Identify B and **ii** suggest a reagent for this reaction. *(2 marks)*

   c. **i** Give the formula of C, and **ii** describe the appearance of $[Cu(NH_3)_6]^{2+}(aq)$. *(2 marks)*

   d. **i** What reagent converts C to $[Cu(NH_3)_6]^{2+}(aq)$? **ii** What type of reaction is this? *(2 marks)*

   e. What is the oxidation state of copper throughout these reactions? *(1 mark)*

25.1 Naming organic compounds

Specification reference: 3.3.1

The IUPAC system

The IUPAC system for naming organic compounds is based on a **root** giving the length of the longest hydrocarbon chain with suffixes (and sometimes prefixes) indicating the **functional groups**. Locants are numbers showing where on the longest chain the functional groups (or hydrocarbon side chains) are found. Table 1 includes all the functional groups you will meet.

> **Synoptic link**
>
> The IUPAC system was introduced in Topic 11.1, Carbon compounds.

▼ **Table 1** *The suffixes and prefixes of some functional groups*

Family	Formula	Suffix	Prefix	Example
alkenes	$RCH=CH_2$	-ene		propene, CH_3CHCH_2
alkynes	$RC\equiv CH$	-yne		propyne, CH_3CCH
halogenoalkanes	R—X (X is F, Cl, Br, or I)		halo- (fluoro-, chloro-, bromo-, iodo-)	chloromethane, CH_3Cl
carboxylic acids	RCOOH	-oic acid		ethanoic acid, CH_3COOH
anhydrides	RCOOCOR'	-anhydride		ethanoic anhydride, $CH_3COOCOCH_3$
esters	RCOOR'	-oate (Esters are named from their parent alcohol and acid, so propyl ethanoate is derived from propanol and ethanoic acid.)		propyl ethanoate, $CH_3COOC_3H_7$
acyl chlorides	RCOCl	-oyl chloride		ethanoyl chloride, CH_3COCl
amides	$RCONH_2$	-amide		ethanamide, CH_3CONH_2
nitriles	$RC\equiv N$	-nitrile		ethanenitrile, $CH_3C\equiv N$
aldehydes	RCHO	-al		ethanal, CH_3CHO
ketones	RCOR'	-one		propanone, CH_3COCH_3
alcohols	ROH	-ol	hydroxy-	ethanol, C_2H_5OH; 2-hydroxyethanal, $HOCH_2CHO$
amines	RNH_2	-amine	amino-	ethylamine, $CH_3CH_2NH_2$
arenes	C_6H_5R			methylbenzene, $C_6H_5CH_3$

Summary questions

1 Name the following:

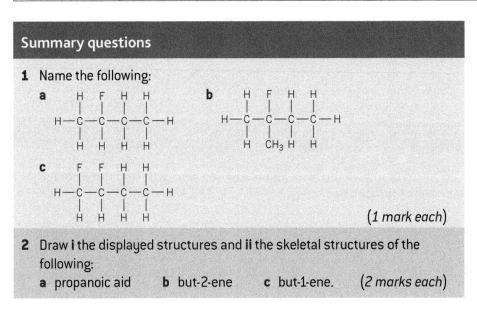

(*1 mark each*)

2 Draw **i** the displayed structures and **ii** the skeletal structures of the following:

   a propanoic aid   **b** but-2-ene   **c** but-1-ene.   (*2 marks each*)

25.2 Optical isomerism

Specification reference: 3.3.7

Key term

Optical isomers: Pairs of molecules that are non-superimposable mirror images.

Synoptic link

The other form of stereoisomerism is E-Z isomerism, covered in Topic 11.3, Isomerism.

Optical isomerism is a particular form of **stereoisomerism**.

Chirality

Optical isomerism occurs when two molecules have the same formula but are non-superimposable mirror images of each other.

Such molecules are described as **chiral** (handed).

Optical isomers have a carbon atom that is bonded to four different atoms or groups of atoms.

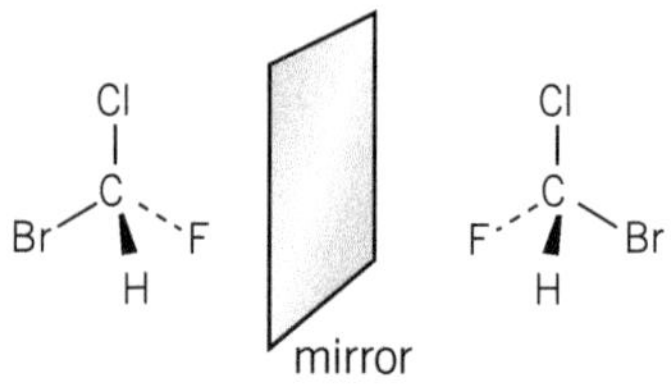

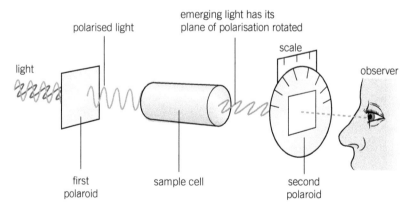

▲ **Figure 1** *The optical isomers of bromochlorofluoromethane*

This atom is called the **chiral centre** and is usually indicated with an asterisk.

It is much easier to see the chiral centre in the displayed formula, which shows all bonds and all atoms. For example the compound pentan-2-ol has the structural formula $CH_3CH_2CH_2CH(OH)CH_3$, the skeletal formula and the displayed formula.

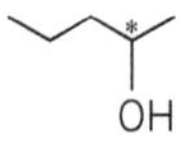

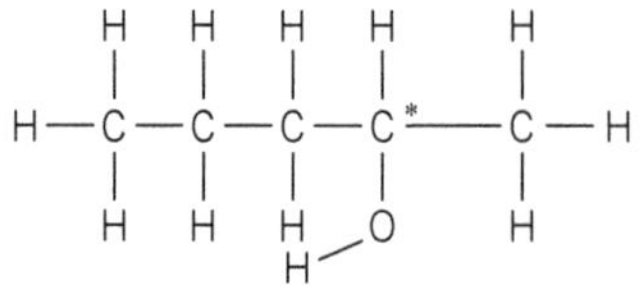

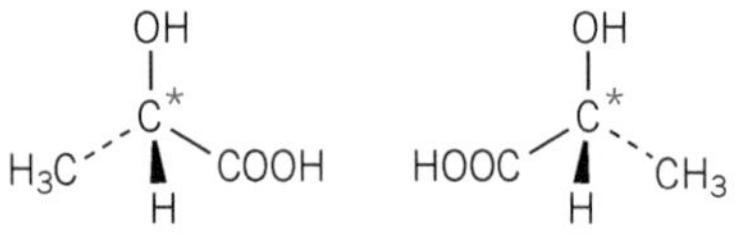

▲ **Figure 2** *The optical isomers of 2-hydroxypropanoic acid*

Optical isomers can *only* be distinguished on paper using 3-D conventions, for example, wedge bonds coming out of and dotted bonds going into the paper as in Figure 1.

Each isomer rotates the plane of polarisation of polarised light in opposite directions, see Figure 2.

▲ **Figure 3** *Measuring the rotation of the plane of polarisation of polarised light*

The + isomer rotates the plane of polarisation clockwise and the − isomer rotates it anticlockwise.

Summary questions

1. Which of the following molecules show optical isomerism? $CHFCl_2$; CF_3Cl; CF_2Cl_2; $CHFClBr$. Explain. *(2 marks)*

2. Mark the chiral centre on the following molecule and draw its optical isomer using the same notation. *(2 marks)*

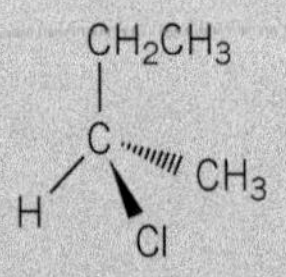

3. α-Amino acids have the formula $RHC(CO_2)(NH_2)$ where R may be H, CH_3, C_2H_5, etc. Give the formula of the *only* α-amino acid that is *not* chiral. Explain your answer. *(2 marks)*

25.3 Synthesis of optically active compounds

Specification reference: 3.3.7

Most chemical reactions that produce optically active compounds produce a 50:50 mixture of the two isomers. This is called a **racemate** (pronounced rass-em-ate) or a **racemic** mixture. The mixture is not optically active because the optical rotations of the two isomers are in opposite directions and cancel each other out. An example of this is the synthesis of 2-hydroxypropanoic acid (lactic acid) from ethanal.

> **Key term**
>
> **Racemate:** A 50:50 mixture of a pair of optical isomers.

Synthesis of 2-hydroxypropanoic acid (lactic acid) from ethanal

The ethanal molecule is flat and therefore not optically active.

Step 1: addition of hydrogen cyanide

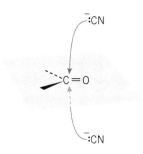

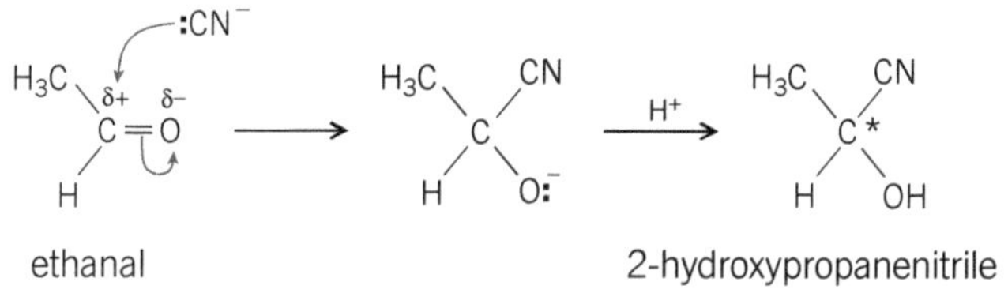

▲ **Figure 1** *The :CN⁻ ion may attack from above or below the molecule*

The key step is the attack of :CN⁻ on $C^{\delta+}$. This is equally likely to happen from above or below, see Figure 1. Each mode of attack will produce a different optical isomer because tetrahedral products are formed.

Step 2: hydrolysis

The nitrile group is converted to a carboxylic acid.

This does not affect chirality so two optical isomers of 2-hydroxypropanoic acid are formed, see Figure 2.

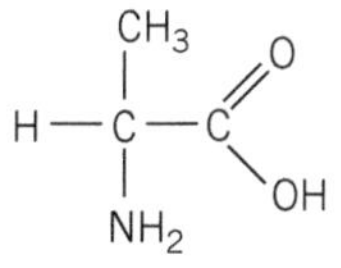

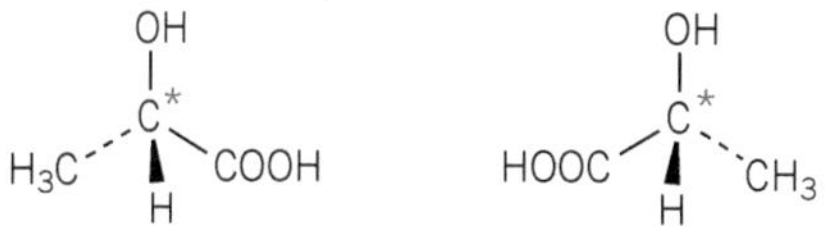

▲ **Figure 2** *The two optical isomers of lactic acid: + on the left and − on the right*

> **Summary questions**
>
> 1 Does the reaction of propanone, below, with hydrogen cyanide produce an optically active product? Explain your answer. *(2 marks)* 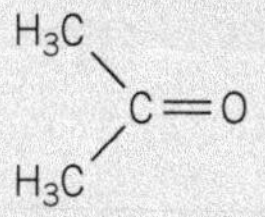
>
> 2 Explain why pairs of optical isomers have identical boiling and melting points. *(2 marks)*
>
> 3 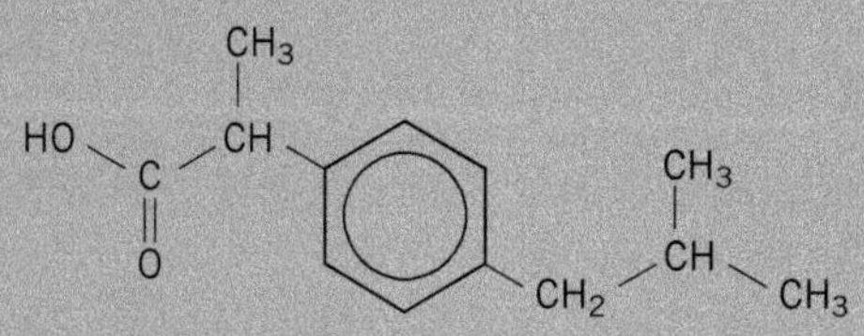
>   a Identify the chiral centre in the structure of the drug ibuprofen above. *(1 mark)*
>   b Suggest a problem that chirality might cause in a drug. *(1 mark)*

Chapter 25 Practice questions

1 The skeletal formula of the drug thalidomide is shown.

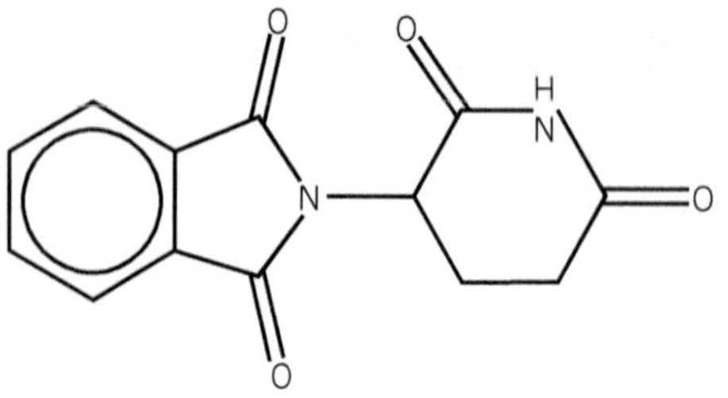

   a Thalidomide is chiral. Explain the term chiral. (2 marks)
   b What problems can chirality cause with a drug molecule? (2 marks)
   c Mark the chiral centre on your drawing. (1 mark)
   d Explain how you identified the chiral centre. Hint - it may help to draw in the hydrogen atoms that are not shown in the skeletal formula. (1 mark)

2 Name and write the shorthand structural formula of the following: (6 marks)

   a 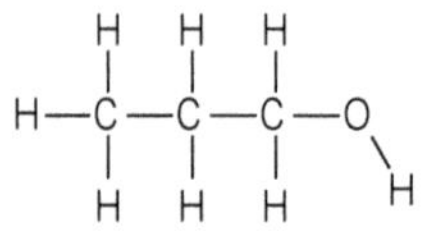
   b 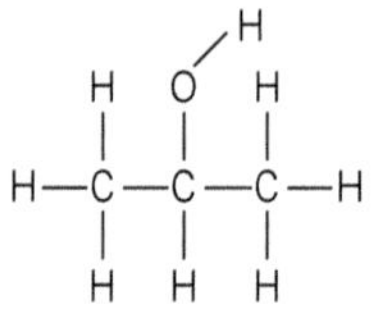
   c

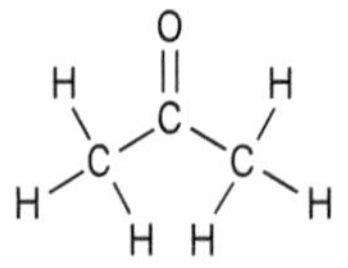

3 The structure of the amino acid alanine is shown.

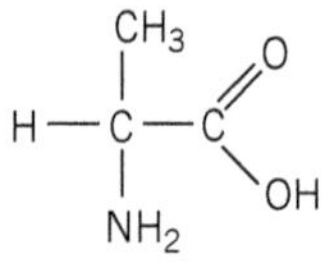

   a How long is the longest unbranched chain of carbon atoms? (1 mark)
   b What is the root of the systematic (IUPAC) name of alanine? (1 mark)
   c Where on the carbon chain is the $-NH_2$ (amino) group? (1 mark)
   d State the systematic name of alanine. (1 mark)
   e What type of isomerism does alanine show? (1 mark)
   f How may these isomers be distinguished from each other? (1 mark)

26.1 Introduction to aldehydes and ketones
Specification reference: 3.3.8

The carbonyl group

The carbonyl group is a carbon–oxygen double bond, C=O.

It is found in aldehydes, RCHO, and ketones, RR'CO.

The simplest aldehyde is methanal,

$$\text{H}\diagdown\text{C}=\text{O} \diagup \text{H}$$

where R=H, followed by ethanal, H_3C—CH=O, where R=CH_3.

The simplest ketone is propanone H_3C—CO—CH_3.

Naming carbonyl compounds

Aldehydes have the suffix –al and ketones the suffix –one.

Aldehydes can only occur at the end of a chain.

Ketones may need a locant to indicate where the C=O group is in a chain, for example, pentan-2-one and pentan-3-one.

Physical properties

The carbonyl group is polarised $C^{\delta+}=O^{\delta-}$. So there are strong permanent dipole–dipole forces between the molecules of aldehydes and ketones.

The reactivity of carbonyl compounds

Because of the polarity of the carbonyl group, **nucleophiles** can attack the $C^{\delta+}$.

The C=O is unsaturated so addition reactions are possible.

The typical reactions of carbonyl compounds are **nucleophilic addition reactions**.

Carbonyl compounds can also be oxidised or reduced by suitable oxidising agents or reducing agents.

Key terms

Nucleophiles: Ions or molecules with a negative charge or a partially negatively charged area that can attack a positively charged area in another molecule and form a bond with it via a lone pair of electrons.

Unsaturated: A compound to which more hydrogen can be added. Unsaturated compounds have double bonds.

Summary questions

1. Which carbonyl compound is most likely to be a gas at room temperature? Explain. *(2 marks)*

2. Explain why there are hydrogen bonds between propanone molecules and water molecules but not between propanone molecules themselves. *(2 marks)*

3. a Draw the structural formula of pentan-3-one. *(1 mark)*
   b State the IUPAC name of the compound shown. *(1 mark)*

26.2 Reactions of the carbonyl group in aldehydes and ketones

Specification reference: 3.3.8

The typical reactions of the carbonyl group are nucleophilic addition and redox.

Nucleophilic addition reactions

The general reaction is

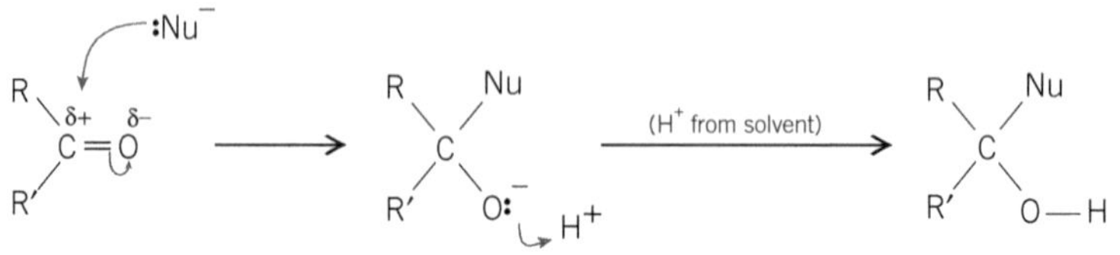

> **Revision tip**
> It is best to remember the general reaction and apply it to different nucleophiles rather than to memorise all the reactions.

> **Revision tip**
> 'Curly arrows' show the movement of pairs of electrons. They move from negatively charged areas towards positively charged areas on molecules.

The nucleophile adds on to the $C^{\delta+}$, and a hydrogen atom (from the solvent) adds on to the $O^{\delta-}$. In the reaction with CN^- ions the products, called hydroxynitriles, have one extra carbon atom than the starting materials. Hydroxynitriles can be easily converted into other functional groups.

Redox reactions

Aldehydes can be oxidised to carboxylic acids and reduced to primary alcohols. Ketones cannot be oxidised easily but can be reduced to secondary alcohols.

> **Revision tip**
> The reagent used to generate CN^- ions is a mixture of KCN and acid. The CN^- ion is very toxic.

Oxidation

The oxidising agent is acidified potassium dichromate solution ($H^+/Cr_2O_7^{2-}$). This is yellow but it is reduced to green Cr^{3+} ions as the reaction proceeds. For example, $RCHO + [O] \rightarrow RCOOH$

Reduction

Sodium tetrahydridoborate(III) (sodium borohydride, $NaBH_4$) is used in aqueous solution to reduce aldehydes and ketones to alcohols. It generates the nucleophile $:H^-$, and the mechanism of the reaction is the same as nucleophilic addition. For an aldehyde:

> **Revision tip**
> It is acceptable to use [O] to represent an oxidising agent, but your equation *must* balance.

> **Revision tip**
> It is acceptable to use [H] to represent a reducing agent, but your equation *must* balance.

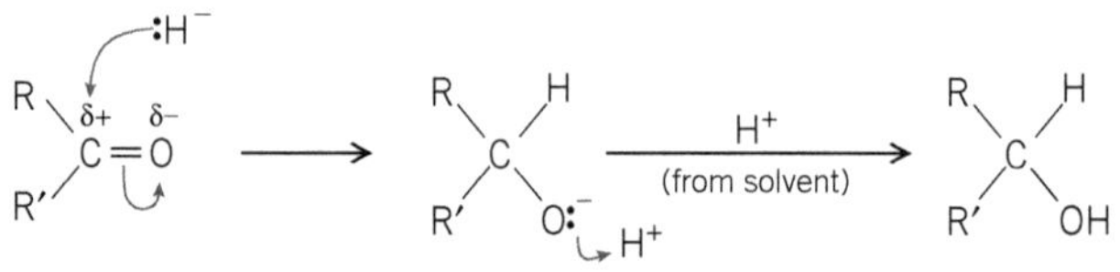

Using [H] to represent the reducing agent the equation is:

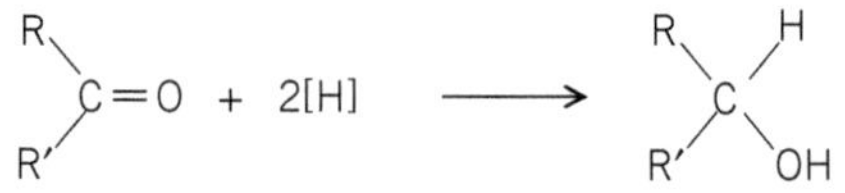

Distinguishing aldehydes from ketones

Two tests may be used. Warming an aldehyde with Tollens' reagent, $Ag(NH_3)_2^+$, forms a silver mirror on the test tube as Ag^+ is reduced to Ag. Ketones give no reaction. Warming an aldehyde with Fehling's, or Benedict's, solution turns the blue solution brick red as the Cu^{2+} ions they contain are reduced to Cu^+. Ketones give no reaction. Both depend on the fact that aldehydes are easily oxidised and ketones are not.

> **Summary questions**
>
> 1. Explain why $:H^-$ is a nucleophile. *(1 mark)*
>
> 2. Suggest why tetrahydridoborate(III) will not reduce C=C. *(1 mark)*
>
> 3. Write the mechanism for the reduction of a ketone with $:H^-$. *(3 marks)*

142

26.3 Carboxylic acids and esters
26.4 Reactions of carboxylic acids and esters
Specification reference: 3.3.9

Carboxylic acids have the general formula RCOOH and esters RCOOR'.

A carboxylic acid   An ester

They both contain the carbonyl group C=O whose properties are modified when adjacent to an –OH group or an –OR group. Esters are derivatives of carboxylic acids where an R' group replaces the hydrogen atom of the OH group of the parent acid.

Naming carboxylic acids

Carboxylic acids have the suffix –oic acid, for example CH_3COOH is ethanoic acid. The carboxylic acid can only occur at the end of a chain so locants are not needed. If there are other groups attached to the main chain we count from the carbon of the carboxylic acid.

Naming esters

Esters are named from the parent acid so all esters based on ethanoic acid are called ethanoates. The name of the R group is placed first, see examples.

methyl ethanoate   *ethyl methanoate*

The reactivity of carboxylic acids

Carboxylic acids are polarised as shown:

- The carbonyl carbon is more strongly polarised $\delta+$ than in aldehydes and ketones due to the second oxygen atom bonded to it. So it is easily attacked by nucleophiles.
- The $O^{\delta-}$ of the carbonyl group may be attacked by positively charged reagents such as H^+.
- The $H^{\delta+}$ may be lost as H^+ ion (a proton). In this case the compound is acting as an acid.

Acidic behaviour

Loss of a proton results in a **carboxylate ion** in which the negative charge is spread over three atoms.

This **delocalisation** makes the ions particularly stable, so carboxylic acids are acidic.

Carboxylic acids are weak acids. They will react with sodium hydrogencarbonate to give off carbon dioxide as well as the usual reactions of acids to form salts.

$RCOOH(aq) + NaHCO_3(aq) \rightarrow RCOO^- Na^+(aq) + CO_2(g) + H_2O(l)$

Formation of esters

Carboxylic acids react with alcohols, using a strong acid catalyst, to form esters. This is a reversible reaction.

> **Revision tip**
> Make sure you can write balanced equations for the reaction of a carboxylic acid with an alkali, a metal, a metal oxide, and a metal carbonate.

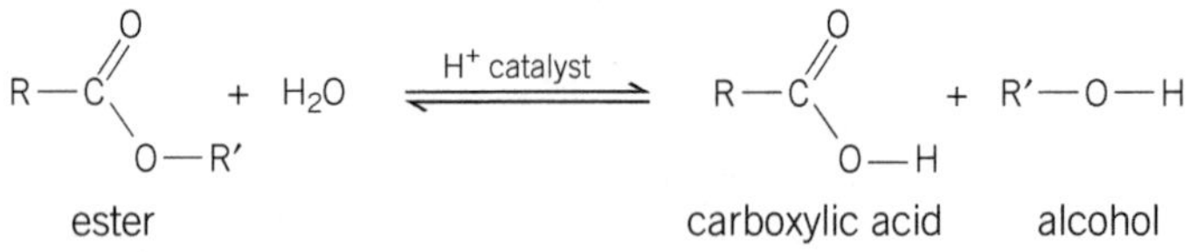

Compounds containing the carbonyl group

$$R-\underset{O-R'}{\overset{O}{\overset{\|}{C}}} + H_2O \underset{}{\overset{H^+ \text{ catalyst}}{\rightleftharpoons}} R-\underset{O-H}{\overset{O}{\overset{\|}{C}}} + R'-O-H$$

ester → carboxylic acid + alcohol

The reverse of this reaction is called acid **hydrolysis** of the ester.

Hydrolysis of esters is also catalysed by bases such as sodium hydroxide. This produces the salt of the acid which removes the acid from the reaction mixture so this reaction goes to completion.

> **Revision tip**
> Catalysts affect the rate of reversible reactions but not the position of equilibrium.

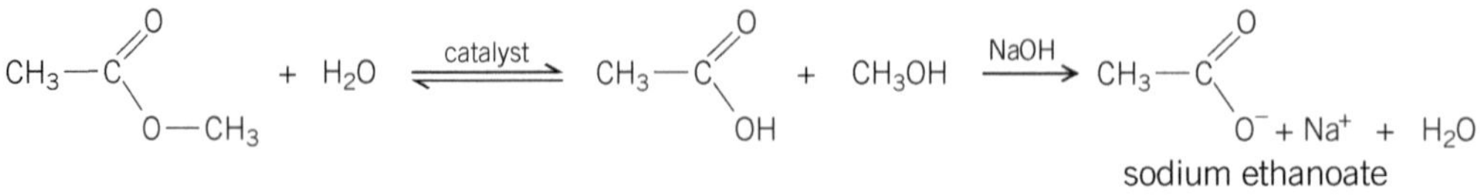

$$CH_3-\underset{O-CH_3}{\overset{O}{\overset{\|}{C}}} + H_2O \overset{\text{catalyst}}{\rightleftharpoons} CH_3-\underset{OH}{\overset{O}{\overset{\|}{C}}} + CH_3OH \overset{NaOH}{\longrightarrow} CH_3-\underset{O^- + Na^+}{\overset{O}{\overset{\|}{C}}} + H_2O$$

sodium ethanoate

The mechanism for this reaction can be explained using 'curly arrows'.

Uses of esters

Common uses of esters include solvents, plasticisers, perfumes, and food flavourings.

Vegetable oils and animal fats are esters of propane-1,2,3-triol (glycerol) with long-chain carboxylic acids. They can be hydrolysed by alkalis to make soaps, see Figure 1.

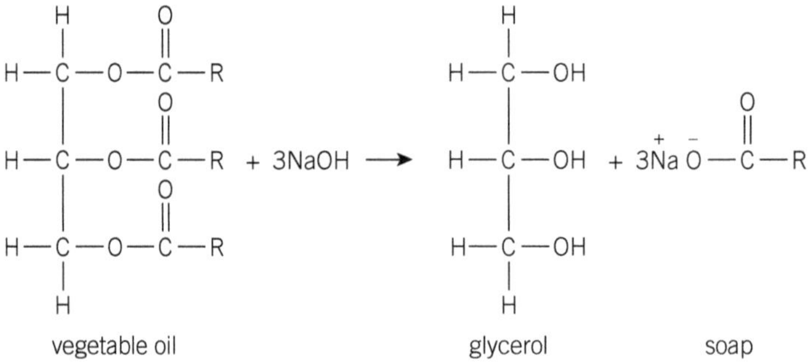

▲ **Figure 1** *The hydrolysis of vegetable oil. R is a long hydrocarbon chain, approximately C_{16}*

Biodiesel is produced by reacting vegetable oils with methanol in the presence of a catalyst to form a mixture of methyl esters of the carboxylic acids that make up the oils.

> ### Go further: Printer inks
>
> Ink jet printer inks need two contrasting properties. They need to be water-soluble while in the cartridge so that they can be sprayed onto the paper, and they need to be insoluble when on the paper so that they do not run. This dilemma is solved by the following equilibrium:
>
> $RCOO^-(aq) + H^+(aq) \rightleftharpoons RCOOH(s)$
>
> RCOOH represents a long-chain carboxylic acid which is coloured and not water-soluble. It is present in the cartridge as its sodium salt, RCOONa, which is water-soluble.
>
> **a** Write an equation for the dissociation of RCOONa into ions.
> Paper is somewhat acidic – that is, it contains H^+ ions.
> **b** What happens to the following equilibrium when $RCOO^-$ ions are in contact with the paper?
> $RCOO^-(aq) + H^+(aq) \rightleftharpoons RCOOH(s)$
> **c** Explain why a long-chain carboxylic acid is insoluble in water while its sodium salt is soluble.

> **Summary questions**
>
> 1 Draw the structural formula of ethyl ethanoate. *(1 mark)*
>
> 2 Name
>
> $H-\underset{H}{\overset{Br}{\overset{|}{C}}}-\underset{H}{\overset{H}{\overset{|}{C}}}-\underset{OH}{\overset{O}{\overset{\|}{C}}}$
>
> *(1 mark)*
>
> 3 What acid and what alcohol would react together to give methyl propanoate? *(2 marks)*
>
> 4 What would be the products of hydrolysis of ethyl butanoate catalysed by **a** hydrochloric acid, **b** sodium hydroxide? *(4 marks)*

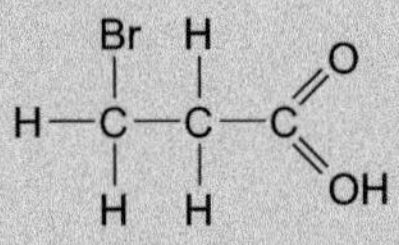

26.5 Acylation

Specification reference: 3.3.9

Acylation is the substitution of the **acyl group**, $R-C\overset{O}{\underset{}{\diagdown}}$ into another molecule.

Acid derivatives contain the acyl group and have the general formula:

$$R-C\overset{O}{\underset{Z}{\diagdown}}$$

Acid derivatives

Acid derivatives are summarised in Table 1.

▼ **Table 1** *Some acid derivatives*

−Z	Name	General formula	Example
−OR'	ester	RCOOR'	ethyl ethanoate $CH_3COOC_2H_5$
−Cl	acid chloride	RCOCl	ethanoyl chloride CH_3COCl
−OCOR'	acid anhydride	RCOOCOR'	ethanoic anhydride $CH_3COOCOCH_3$
−NH$_2$	amide	RCONH$_2$	ethanamide CH_3CONH_2

Nucleophilic addition–elimination reactions

In all acid derivatives the carbonyl carbon is polarised $C^{\delta+}$ because it is bonded to an electronegative oxygen atom. This $C^{\delta+}$ will be attacked by nucleophiles, followed by loss of: Z^- – the **leaving group**:

$$R-\underset{Z}{\overset{\overset{\delta-}{O}}{C^{\delta+}}} + :Nu^- \longrightarrow R-\underset{Nu}{\overset{O}{C}} + :Z^-$$

> **Revision tip**
> Addition–elimination reactions are so called because the nucleophile is *added* and the leaving group is *eliminated*, so they are substitution reactions.

The speed of the reaction is governed by:

- The size of the δ+ charge of the carbonyl carbon which depends on the electron-attracting power of Z.
- How easily Z is lost (how good a leaving group it is).
- How good the nucleophile is.

Nucleophiles

Typical nucleophiles include

$$:\underset{H}{\overset{H}{\underset{|}{\overset{|}{N}}}}-R \;>\; :\underset{H}{\overset{H}{\underset{|}{\overset{|}{N}}}}-H \;>\; :\underset{R}{\overset{H}{\diagdown\diagup}} \;>\; :\underset{H}{\overset{H}{\diagdown\diagup}}$$

primary amine   ammonia   alcohol   water

> **Revision tip**
> Nucleophiles have a lone pair of electrons which they use to form a bond with $C^{\delta+}$.

Compounds containing the carbonyl group

The reactions are summarised in Table 2.

▼ **Table 2** *Products of the reactions of acid derivatives with nucleophiles*

nucleophile ↓ / acid derivative →	acid chloride R—COCl	anhydride (RCO)₂O
NH_3 ammonia	R—CO—NH₂ amide	R—CO—NH₂ amide
R′—NH₂ amine	R—CO—NHR′ N-substituted amide	R—CO—NHR′ N-substituted amide
R′—OH alcohol	R—CO—O—R′ ester	R—CO—O—R′ ester
H_2O water	R—CO—OH carboxylic acid	R—CO—OH carboxylic acid

(increasing reactivity across; increasing reactivity down)

A typical mechanism is shown for ethanoyl chloride and water.

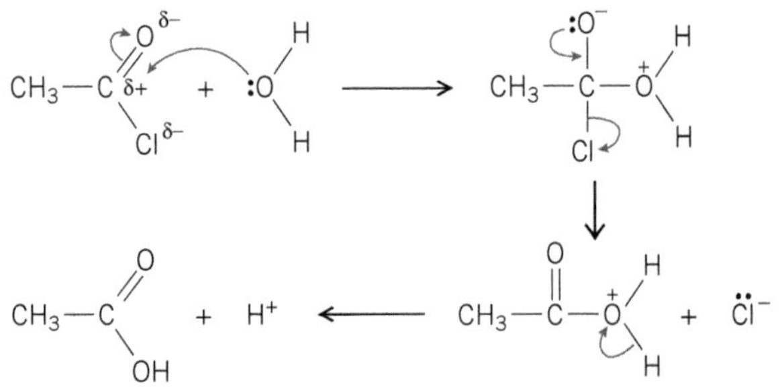

Notice that as the nucleophile (water) is neutral and the leaving group is negatively charged, a H⁺ ion must be lost in the final step.

The overall reaction is

$$CH_3COCl + H_2O \rightarrow CH_3OOH + HCl$$

Aspirin

Aspirin is made by acylating 2-hydroxybenzoic acid with ethanoic anhydride:

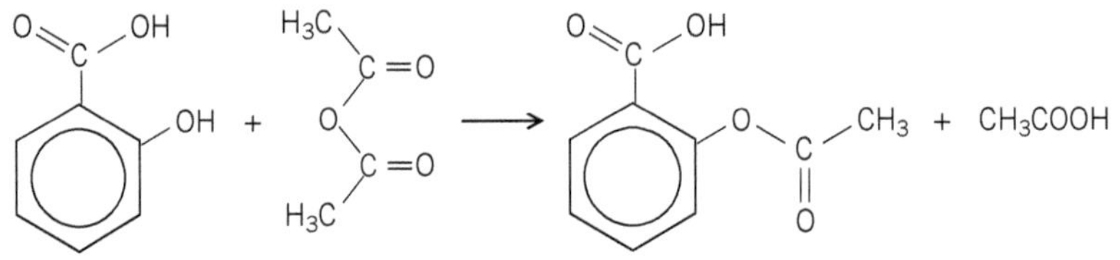

An alternative acylating agent is ethanoyl chloride but the anhydride is used because it is:

- cheaper
- less corrosive
- produces a safer by-product – ethanoic acid rather than hydrochloric acid.

Summary questions

1. Write the balanced equation for the reaction of ethanol with ethanoyl chloride. *(2 marks)*

2. Suggest why :NH₃ is a better nucleophile than :OH₂. *(2 marks)*

3. Calculate the atom economy of the reaction used to prepare aspirin. *(2 marks)*

Chapter 26 Practice questions

1 a State the IUPAC name of the following:

   i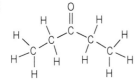

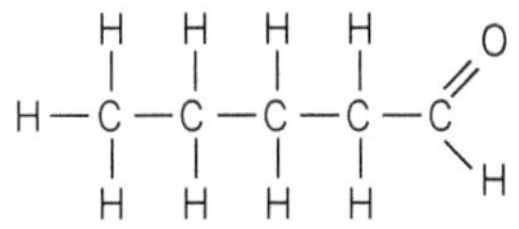

   ii

   iii

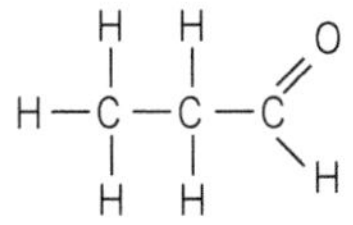

   (3 marks)

   b Which two are a pair of isomers? (1 mark)
   c Which two are aldehydes? (2 marks)
   d Which one is not easily oxidised? (1 mark)
   e Which one can be reduced to a secondary alcohol? (1 mark)
   f Name a reducing agent that would reduce all three. (1 mark)
   g Which two can be oxidised to carboxylic acids? (2 marks)

2 a i Draw the displayed formula of butanone.
     ii Why is no locant needed in its name? (2 marks)

3 a What are the products of the acid hydrolysis of i ethyl methanoate, ii methyl ethanoate? (4 marks)
   b How would the products differ if a catalyst of sodium hydroxide were used? (1 mark)
   c Write an equation for i the acid hydrolysis of ethyl methanoate and ii the hydrolysis of methyl ethanoate catalysed by sodium hydroxide. (4 marks)

4 a Write a balanced equation for the reduction of ethanal to ethanol by sodium tetrahydridoborate(III) (sodium borohydride) using [H] to represent the reducing agent. (1 mark)
   b Explain why this reducing agent will reduce compounds containing C=O but not those containing C=C. (2 marks)

27.1 Introduction to arenes

Specification reference: 3.3.10

Bonding and structure of benzene

Aromatic compounds are based on benzene, C_6H_6, which is a regular, flat hexagonal ring of carbon atoms, each one bonded to a single hydrogen atom. Although it is unsaturated, it does not usually undergo addition reactions. The bond lengths are all the same and are intermediate between C–C and C=C. Benzene has the special symbol

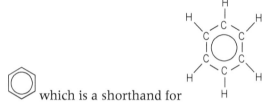

which is a shorthand for

The skeleton of the ring is formed from C–C single bonds but each carbon has a p-orbital containing one electron, Figure 1.

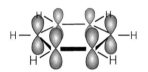

▲ **Figure 1** *The p-orbitals in benzene*

These orbitals overlap to form a **delocalised system** of electrons above and below the ring, see Figure 2.

▲ **Figure 2** *Delocalisation in benzene*

This electron-rich area accounts for much of the chemistry of benzene.

The stability of benzene

Benzene is thermochemically much more stable than would be expected if it were 1,3,5-cyclohexatriene, ⬡.

The experimental enthalpy change of hydrogenation of cyclohexene is $-120\,kJ\,mol^{-1}$.

◯ + H₂ → ◯ $\Delta H = -120\ kJ\ mol^{-1}$

So we would expect that the value for 1,3,5-cyclohexatriene would be three times this, that is, $-360\,kJ\,mol^{-1}$.

The experimental value for benzene is $-208\,kJ\,mol^{-1}$ indicating that benzene is $152\,kJ\,mol^{-1}$ more stable than expected, see Figure 3.

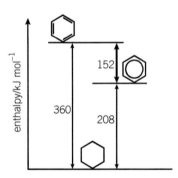

▲ **Figure 3** *Enthalpy diagram for the hydrogenation of 1,3,5-cyclohexatriene, cyclohexane, and benzene*

Common misconception

Do not confuse hydrogenation (reaction with hydrogen) with hydration (reaction with water).

Summary questions

1. Suggest the expected value for the enthalpy of hydrogenation of 1,3-cyclohexadiene. *(1 mark)*

2. The bond length of C–C is 0.154 nm and that of C=C is 0.134 nm. Predict the carbon–carbon bond length in benzene. *(1 mark)*

3. Write a balanced equation for the hydrogenation of benzene. *(1 mark)*

27.2 Arenes – physical properties, naming, and reactivity
27.3 Reactions of arenes
Specification reference: 3.3.10

Benzene is non-polar and mixes well with hydrocarbon solvents but will not dissolve in water. It is a liquid at room temperature.

Naming arenes
Substituted arenes are named as derivatives of benzene with locants to define the positions on the ring where there are two or more of them. Here are some examples:

Methylbenzene   1,2-dichlorobenzene   1,4-dichlorobenzene

Some exceptions include:

phenol   benzoic acid (benzenecarboxylic acid)

> **Key term**
> **Polar:** A molecule with areas of δ+ and δ− partial charges.

> **Revision tip**
> When naming organic compounds, there are commas between strings of numbers and hyphens between numbers and words.

The reactivity of arenes
- The aromatic system is very stable and is rarely destroyed in reactions. So the typical reactions of arenes are substitutions.
- The aromatic ring has a high electron density and is attacked by electrophiles.

So the reactions of arenes are typically **electrophilic substitutions**.

Electrophilic substitution reactions of arenes
Mechanism
The general mechanism is

> **Revision tip**
> A H^+ ion is a proton – it has no electrons.

This is the general equation for electrophilic substitution. Notice that a H^+ ion is lost from the benzene ring and replaced by the electrophile El^+.

Aromatic chemistry

> **Synoptic link**
> Here sulfuric acid is the stronger acid and donates a H⁺ ion to nitric acid, see Topic 21.1, Defining an acid.

Nitration

The electrophile is NO_2^+, produced in a mixture of concentrated nitric and sulfuric acids:

$$H_2SO_4 + HNO_3 \rightarrow H_2NO_3^+ + HSO_4^- \rightarrow NO_2^+ + HSO_4^- + H_2O$$

The product is nitrobenzene. Further reaction produces small amounts of 1,2-dinitrobenzene and 1,4-dinitrobenzene.

Nitrobenzene can be reduced using tin and hydrochloric acid to phenylamine. Phenylamine and other substituted aromatic amines are important starting materials for dyestuffs.

Friedel–Crafts acylation

The electrophile is RCO^+, generated from an acyl chloride and an aluminium chloride catalyst:

$$RCOCl + AlCl_3 \rightarrow RCO^+ + AlCl_4^-$$

The product is an acyl-substituted arene.

The balanced equation is

benzene + RCOCl → acyl-substituted arene + HCl

Summary questions

1. Name and draw the structure of the third isomer of 1,2-dichlorobenzene and 1,4-dichlorobenzene. *(2 marks)*

2. Write the equation for **a** the nitration of benzene with the electrophile NO_2^+. **b** the reduction of nitrobenzene to phenylamine by hydrogen. *(4 marks)*

3. Explain how the catalyst $AlCl_3$ is regenerated in the reaction of RCO^+ with benzene. *(2 marks)*

4. What is the functional group of (phenyl methyl ketone), which is produced by the reaction of CH_3COCl with benzene? *(1 mark)*

Chapter 27 Practice questions

1 **a** State the IUPAC name of the following:

   i (bromobenzene structure)

   ii (ethylbenzene structure)

   iii (1,4-dimethylbenzene structure with CH₃ and H₃C)

   (3 marks)

   b Which two are a pair of isomers? (1 mark)

   c Draw the structure of 1,2-dimethylbenzene. (1 mark)

2 Explain why the reactions of aromatic compounds are almost always electrophilic substitution reactions. (2 marks)

3 Benzene can be converted to nitrobenzene by an electrophilic substitution reaction with a mixture of concentrated nitric and sulfuric acids.

   a What is the electrophile in this reaction? (1 mark)

   b Write an equation to show how this electrophile is formed in the acid mixture. (2 marks)

   c Write a balanced equation for the substitution reaction between this electrophile and benzene. (2 marks)

   d Nitrobenzene can be converted to phenylamine by reduction.

   i Write a balanced symbol equation for this reaction using [H] to represent the reducing agent. (2 marks)

   ii What reducing agent is actually used? (1 mark)

28.1 Introduction to amines

Specification reference: 3.3.11

The structure of amines

Amines are classified as primary (1°), secondary (2°) or tertiary (3°) depending on how many hydrogen atoms have been replaced by organic groups.

H—N̈—H   H—N̈—H   H—N̈—R'   R"—N̈—R'
| | | |
H   R   R   R

ammonia   primary amine   secondary amine   tertiary amine

Naming amines

Amines are named using the suffix –amine, so:

CH_3NH_2 is methylamine and $C_2H_5NH_2$ is ethylamine;

$(CH_3)_2NH$ is dimethylamine and $(C_2H_5)_3N$ is triethylamine.

Different substituents are written in alphabetical order, so

$(C_6H_5)(CH_3)NH$ is methylphenylamine.

Physical properties of amines

Amines are pyramidal-shaped molecules with bond angles of approximately 107°. They have a lone pair of electrons which squeezes down the bond angles from the ideal 109.5° of a tetrahedron.

Reactivity of amines

The reactivity of amines is governed by the lone pair of electrons. This means that they are **nucleophiles** and **bases**.

> **Key term**
>
> **Amines:** Derivatives of ammonia, NH_3, in which one or more hydrogen atoms has been replaced by an organic group.

> **Revision tip**
>
> Organic groups are classified as alkyl or aryl. Alkyl groups are based on alkanes (e.g., C_2H_5, ethyl) and aryl groups are based on aromatic compounds such as benzene (e.g., C_6H_5, phenyl).

> **Revision tip**
>
> In amines, 1°, 2°, and 3° refer to the number of organic groups attached to the *nitrogen* atom. In alcohols, 1°, 2°, and 3° refer to the number of organic groups attached to the *carbon bonded to the OH group* (see Topic 15.1, Alcohols – introduction).

> **Synoptic link**
>
> Revise the factors that determine the shapes of molecules in Topic 3.6, The shapes of molecules and ions.

> **Revision tip**
>
> Nucleophiles use their lone pair to attack $C^{\delta+}$. Bases use their lone pair to accept a H^+ ion (a proton).

> **Summary questions**
>
> 1 a Classify [C₆H₅–NH–CH₃ structure shown] as a primary, secondary, or tertiary amine. Explain your answer. *(2 marks)*
>   b Name the amine in **a**. *(1 mark)*
>
> 2 Draw the structural formulae of the three isomers of the compound in question 1 which are primary amines. *(3 marks)*

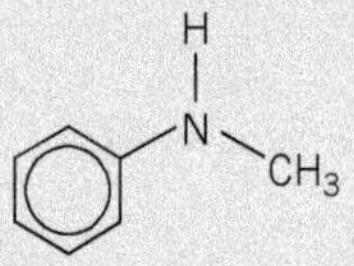

28.2 The properties of amines as bases
Specification reference: 3.3.11

Amines as bases

Amines are basic because they have a lone pair of electrons on the nitrogen atom that can form a bond with a H⁺ ion, often termed 'accepting a proton'.

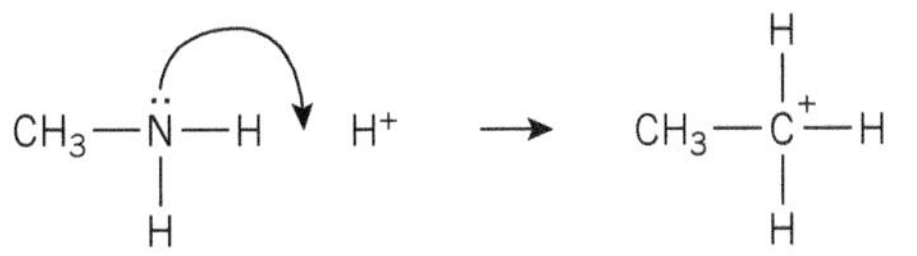

Revision tip
The bond that is formed is called a co-ordinate bond because both the electrons in the bond come from the same atom.

Reactions of amines as bases

Amines react with acids to form ionic salts, for example:

$C_2H_5NH_2 + H^+ + Cl^- \longrightarrow C_2H_5NH_3^+ + Cl^-$
ethylamine   ethylammonium chloride

Revision tip
Look back at Topic 21.1, Defining an acid, to revise the idea of a Brønsted–Lowry base.

The products are named in the same way as salts of ammonia, e.g., $C_6H_5NH_3Cl$ is phenylammonium chloride, CH_3NH_3Cl is methylammonium chloride.

Adding a strong base will remove the proton and re-form the amine.

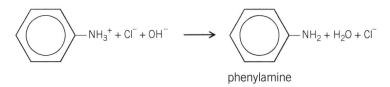

phenylamine

The base strengths of amines

Base strengths of alkyl amines have the pattern:

Secondary amine > primary amine > ammonia > aryl amine

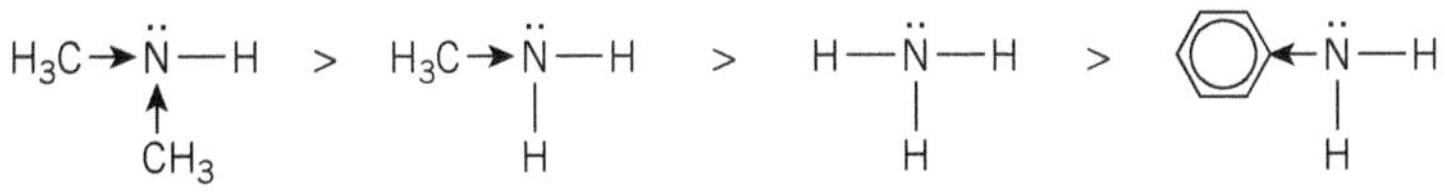

This is because of the **inductive effect** of alkyl groups, which release electrons, making the lone pair more available for donation. Aryl amines are weaker bases than ammonia because the lone pair of electrons is delocalised onto the aromatic ring.

Revision tip
The inductive effect is shown by an arrow along the bond in the direction of electron release.

Summary questions

1. **a** State the formula of dimethylammonium chloride. *(1 mark)*
   b Name and give the formula of an isomer of dimethylammonium chloride. *(1 mark)*
   c Would you expect dimethylammonium chloride to dissolve in water? Explain your answer. *(2 marks)*

2. Dimethylamine and ethylamine are isomers. **a** Give the formula of each.
   b Which is the stronger base? Explain your answer. *(4 marks)*

28.3 Amines as nucleophiles and their synthesis

Specification reference: 3.3.11

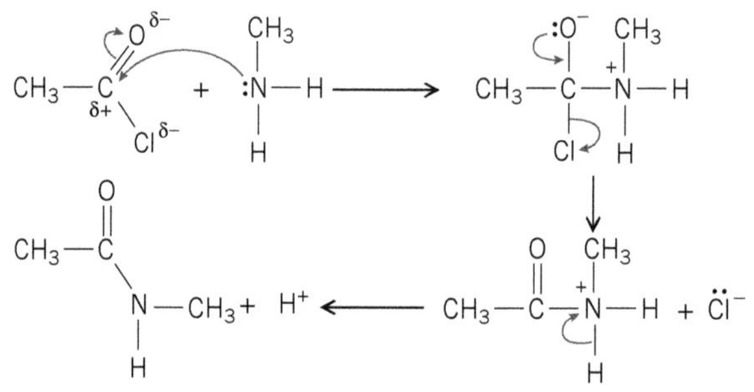

> **Key term**
>
> **Nucleophiles:** Chemical species with a lone pair of electrons with which they can form a bond with $C^{\delta+}$.

> **Synoptic link**
>
> The mechanism of the nucleophilic substitution reaction between ammonia and halogenoalkanes is discussed in Topic 13.2, Nucleophilic substitution in halogenoalkanes.

> **Synoptic link**
>
> Addition–elimination reactions of amines with acid chlorides and acid anhydrides are covered in Topic 26.5, Acylation.

> **Synoptic link**
>
> Nitriles can be prepared from halogenoalkanes, see Topic 13.2, Nucleophilic substitution in halogenoalkanes.

> **Synoptic link**
>
> The nitration of benzene is an electrophilic substitution. It is covered in Topic 27.3, Reactions of arenes.

Nucleophilic substitution reactions of amines

Reactions of ammonia and amines with halogenoalkanes

Ammonia and amines all have a lone pair on the nitrogen atom with which they can attack compounds with a $C^{\delta+}$. This makes them nucleophiles.

The reaction of ammonia with halogenoalkanes is a nucleophilic substitution reaction. It takes place in ethanol with excess ammonia under pressure.

The first step is: $RCH_2X + 2NH_3 \rightarrow RCH_2NH_2 + NH_4X$

The product is a primary amine which is also a nucleophile.

With excess halogenoalkane, further nucleophilic substitutions occur to form secondary and tertiary amines and eventually quaternary ammonium salts:

$RCH_2NH_2 + RCH_2X \rightarrow HX + (RCH_2)_2NH$ (secondary amine)

$(RCH_2)_2NH + RCH_2X \rightarrow HX + (RCH_2)_3N$ (tertiary amine)

$(RCH_2)_3N + RCH_2X \rightarrow X^- + (RCH_2)_4N^+$ (a quaternary ammonium salt)

The formation of amides

Amines will also act as nucleophiles with acid chlorides and acid anhydrides. These are usually described as **addition–elimination reactions**.

An example of the mechanism is:

Preparation of amines

Alkyl amines

Primary amines can be prepared by the reduction of nitriles, $R-C\equiv N$, with hydrogen with a nickel catalyst:

$R-C\equiv N + 2H_2 \rightarrow RCH_2NH_2$

Aromatic amines

Phenylamine and ring-substituted phenylamines can be made by first nitrating the arene and then reducing the nitro-compound that is produced with tin and hydrochloric acid. For example, with benzene:

Amines

benzene + HNO₃ →(conc. H₂SO₄) nitrobenzene + H₂O

nitrobenzene + 6[H] →(Sn/HCl, room temp.) aniline + 2H₂O

Aromatic amines are used in making dyes.

Go further: Explosives

Nitroglycerine is a powerful explosive that can be set off with a small shock. Unlike many explosives, it needs no oxygen from the air to allow it to explode.

CH₂—ONO₂
|
CH—ONO₂
|
CH₂—ONO₂

The products following an explosion are nitrogen, carbon dioxide, water (steam), and oxygen.

a Write a balanced equation for the explosion of nitroglycerine. (Hint, start with 4 molecules of nitroglycerine.)
ΔH for this reaction is −29 000 kJ mol⁻¹, which helps to explain why the reaction is spontaneous.
b How many moles of gas are produced from the four moles of nitroglycerine?
c What factor other than the large negative value of ΔH helps to make this reaction likely to occur?

Summary questions

1 Explain why the reaction of ammonia with an excess of halogenoalkane is *not* a useful way of preparing a secondary amine. *(2 marks)*

2 What is the atom economy of the reaction of a nitrile with hydrogen? Explain. (Hint, no calculation is needed.) *(1 mark)*

3 Explain why secondary and tertiary amines cannot be produced by reduction of nitriles. *(1 mark)*

29.1 Condensation polymers

Specification reference: 3.3.12

Synoptic link

Addition polymers are covered in Topic 14.3. Make sure you understand the difference between these and condensation polymers.

Condensation polymers such as polyesters and polyamides are produced from two different monomers each with two functional groups. These functional groups react together to link the monomers and eliminate a small molecule.

Polyesters

Polyesters are made from a diol and a dicarboxylic acid.

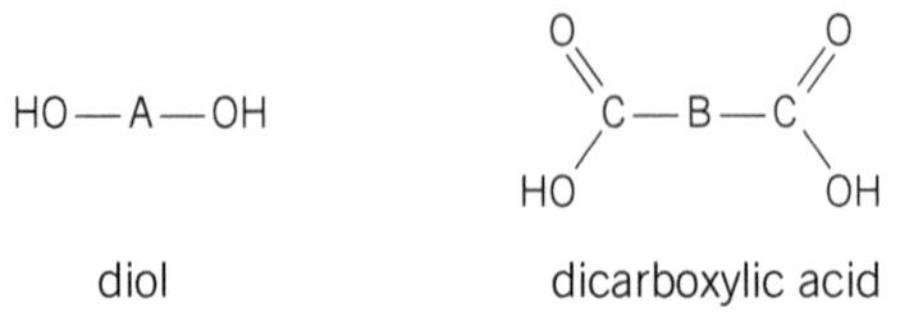

HO—A—OH
O
O
C—B—C
HO
OH
diol
dicarboxylic acid

A and B are unreactive organic groups, for example, CH_2CH_2.

The two monomers react together as shown to eliminate molecules of water and form a long chain (see Figure 1).

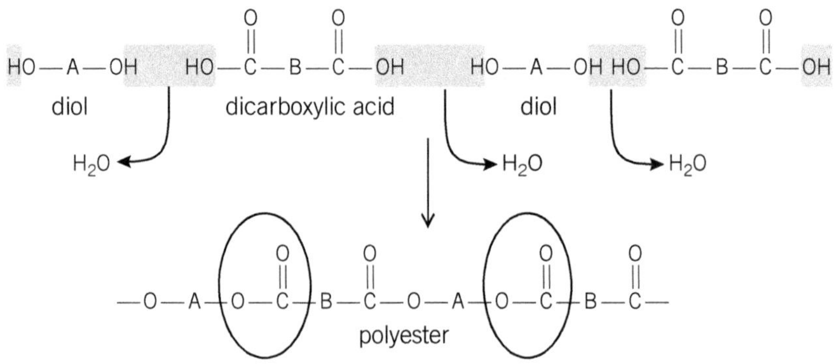

HO—A—OH
HO—C—B—C—OH
HO—A—OH HO—C—B—C—OH
diol
dicarboxylic acid
diol
H2O
H2O
H2O
—O—A—O—C—B—C—O—A—O—C—B—C—
polyester

▲ **Figure 1** *Making a polyester*

The links in the chain, circled, are esters – see Topic 26.4, Reactions of carboxylic acids and esters – and the chain is called a polyester, 'poly' meaning many. The groups A and B can be varied and can tailor the properties of the polymer. In Terylene, for example, A is CH_2CH_2 and B is a benzene ring.

Polyamides

Here the two monomers are a diamine and a dicarboxylic acid (or a diacid chloride).

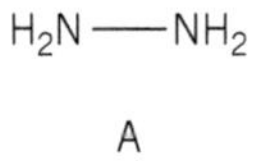

H2N——NH2
A

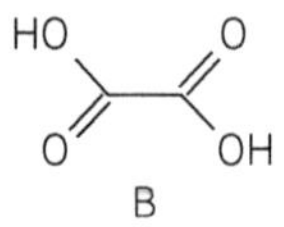

HO
O
O
OH
B

These two monomers also react together to eliminate water and form a chain (see Figure 2).

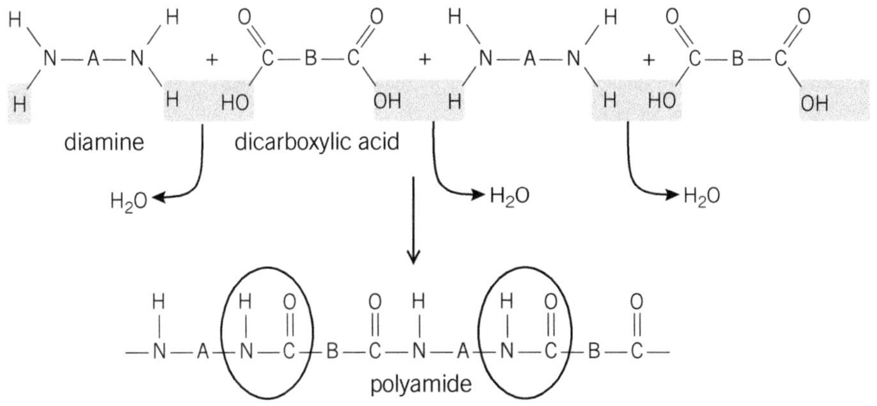

▲ **Figure 2** *Making a polyamide*

The links in the chain are amides, see Topic 26.5, Acylation, and the chain is called a polyamide.

The groups A and B can be varied and can tailor the properties of the polymer. In Nylon 6,6, A is $CH_2CH_2CH_2CH_2CH_2CH_2$ and B is $CH_2CH_2CH_2CH_2$. In Kevlar, both A and B are benzene rings.

Identifying the repeat unit

The repeat unit is found by starting at any point in the polymer chain and stopping when the same group of atoms begins again.

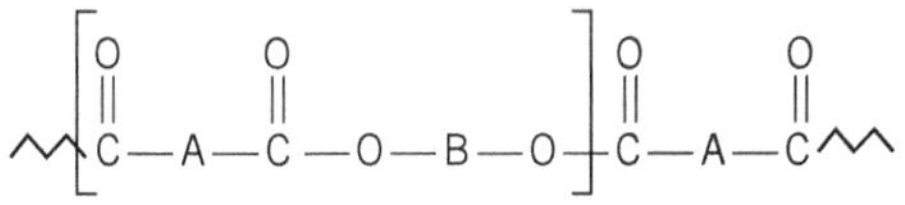

▲ **Figure 3** *The repeat unit is in brackets*

Disposal and degradation of polymers

Condensation polymers are potentially biodegradable and can be broken down by hydrolysis – the reverse of the reaction by which they were made. This reaction is very slow under everyday conditions.

Recycling is possible, for example, fleece garments may be made from the plastics used to make soft drinks bottles.

Summary questions

1. Give the structural formulae of the two monomers for Terylene. *(2 marks)*

2. What small molecule is eliminated if a polyamide is made from a diamine and a diacid chloride? *(1 mark)*

3. Suggest two catalysts for the hydrolysis of a polyester. *(2 marks)*

Chapter 28 and 29 Practice questions

1. **a** Explain what is meant by the terms primary, secondary, and tertiary as applied to amines. *(3 marks)*

   b What feature of an amine molecule means that it can act as a base? *(1 mark)*

   c Amines can also act as nucleophiles. Explain the difference between a base and a nucleophile. *(1 mark)*

2. **a** Name and give the shorthand structural formulae of two compounds from which propylamine, $CH_3CH_2CH_2NH_2$, may be prepared. *(4 marks)*

   b For each case, outline the reaction by which propylamine can be formed, give a suitable reagent and state the type of reaction involved. *(6 marks)*

   c Which of these methods is a more efficient synthesis? Explain your answer. *(2 marks)*

3. Phenylamine, (C₆H₅NH₂), is only sparingly soluble in water and forms oily drops with a characteristic smell on the surface of a beaker of water. On adding hydrochloric acid and shaking, the oily drops and the smell disappear. On adding excess sodium hydroxide the smell and the oily droplets reappear.

   a Explain why phenylamine is only sparingly soluble in water. *(1 mark)*

   b Explain the observations above on adding acid and alkali. *(2 marks)*

4. Explain the difference between an addition polymer and a condensation polymer. *(4 marks)*

5. A section of a polymer molecule is shown.

   ·····C(H)(H)—C(H)(Cl)—C(H)(H)—C(H)(Cl)—C(H)(H)—C(H)(Cl)·····

   a Identify the repeating unit of this polymer. *(1 mark)*

   b Give the name and structural formula of the monomer from which this polymer can be made. *(2 marks)*

   c What type of polymerisation is this? *(1 mark)*

   d A section of a different polymer is shown.

   O—CH₂CH₂—O—C(=O)—C₆H₄—C(=O)—O—CH₂CH₂—O—C(=O)—C₆H₄—C(=O)

   Identify the repeating unit of this polymer. *(2 marks)*

   e Identify **two** monomers from which it may be formed. *(2 marks)*

   f What type of polymer is this? *(1 mark)*

   g Which of the two polymers is more likely to be biodegradable? Explain your answer. *(2 marks)*

6. Both Nylon and proteins are polyamides. Explain how these polymers differ in the structures of their monomers. *(4 marks)*

30.1 Introduction to amino acids
30.2 Peptides, polypeptides, and proteins
Specification reference: 3.3.13

Introduction to amino acids

α-Amino acids (also called 2-amino acids) are the building blocks of proteins. They have the general formula,

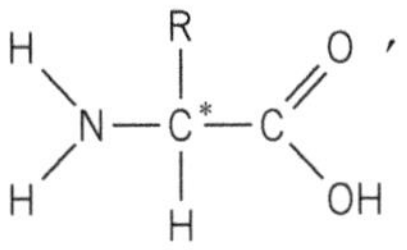

where R can be a variety of organic groups.

Unless R is H, the molecule is chiral. They have a COOH group that is acidic and an NH_2 group that is basic. The chiral centre is marked with an asterisk.

Zwitterions

Amino acids exist as **zwitterions** in which the COOH group has lost a H^+ ion (a proton) and the NH_2 group has gained a H^+ ion (been protonated).

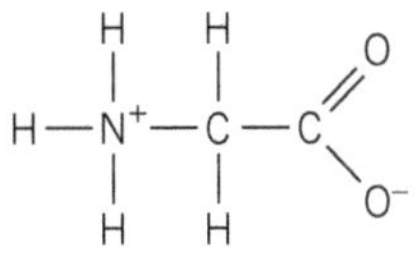

Zwitterions are neutral overall. But have properties similar to ionic compounds – they dissolve in water and have high melting points.

The 20 naturally occurring amino acids are given three-letter names, for example, gly for glycine (R=H), ala for alanine (R=CH_3), etc.

Many amino acids have functional groups as part of their R group. For example, serine (ser) has an OH group, glutamic acid (glu) and aspartic acid (asp) have COOH groups, and lysine (lys) has an NH_2 group.

Peptides, polypeptides, and proteins
The peptide link

Amino acids have two functional groups – an amine and a carboxylic acid – that can react together to form an **amide**. The resulting compound is called a **dipeptide** and this link between amino acids is called a **peptide link**, see Figure 1.

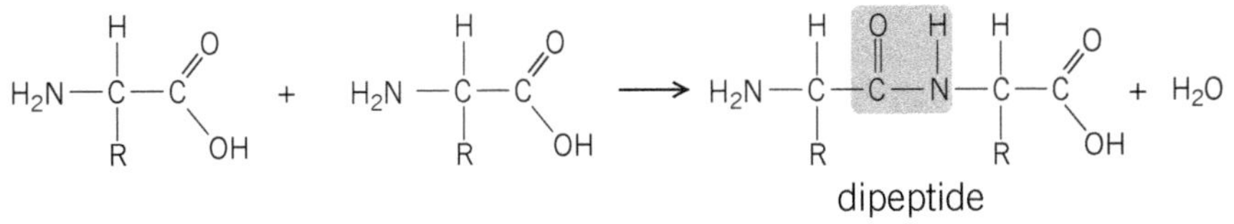

▲ **Figure 1** *Formation of a dipeptide*

Further amino acids can add on to form a chain called a **polypeptide**. Polypeptides longer than around 50 amino acids are called **proteins**.

Hydrolysis of peptides

Boiling a peptide or protein with 6 mol dm⁻³ hydrochloric acid hydrolyses the peptide links and produces a mixture containing all the individual amino acids. The individual amino acids can be identified by **thin-layer chromatography**, see Topic 33.1, Chromatography.

> **Synoptic link**
> Chiral means handed – see Topic 25.2, Optical isomerism. Chiral molecules exist as pairs of optical isomers.

> **Revision tip**
> You can recognise chiral molecules because they have a carbon atom bonded to four different atoms or groups.

> **Key term**
> **Zwitterions:** Ions that have both a positive and a negative charge on the same entity.

> **Key term**
> **Peptide:** A compound made by linking two or more amino acids together.

Amino acids, proteins, and DNA

The structure of proteins

Proteins have three levels of structure.

Primary structure

This is the sequence of amino acids along a protein chain and can be written using the three letter shorthand names for the amino acids, for example, gly–gly–ala–leu–asp, etc. The primary structure is held in place by the covalent bonding of the peptide linkages.

Secondary structure

Protein chains have complex shapes including helixes (spirals) and pleated sheets. These shapes are held in place by three types of bonding:

- Hydrogen bonding between, for example, C=O and NH_2 groups, see Figure 2.
- Ionic bonding between, for example, COO^- groups and NH_3^+ groups.
- Sulfur–sulfur bonds that can form between two cysteine amino acids that have S–H groups in their side chain that can link together.

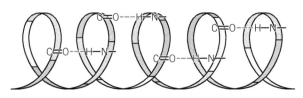

▲ **Figure 2** *A protein helix held together by hydrogen bonding*

Tertiary structure

The helix or pleated sheet can itself be twisted into a three-dimensional shape held by the same types of bonds as the secondary structure. The shapes of proteins are vital for their action as enzymes, see Figure 3.

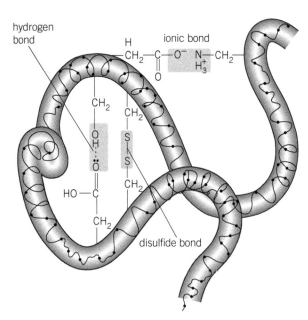

▲ **Figure 3** *Part of a tertiary structure of a helical protein*

Summary questions

1. Name the two functional groups of all amino acids. *(2 marks)*

2. What feature of the NH_2 group of an amino acid allows it to accept a proton? *(1 mark)*

3. Draw the ions formed from an amino acid in **a** a strongly acidic solution, **b** a strongly alkaline solution. *(2 marks)*

4. Draw a hydrogen bond formed between a C=O and an NH_2 group. Show the lone pair involved. *(2 marks)*

5. State the conditions for a hydrogen bond to form. *(2 marks)*

6. Draw the displayed formulae of the amino acids produced by hydrolysis of the dipeptide gly–ala. *(2 marks)*

30.3 Enzymes

Specification reference: 3.3.13

What are enzymes?

Enzymes are large protein molecules, usually globular in shape, that are found in living things. They act as extremely efficient catalysts. They are highly specific – each enzyme catalysing a particular reaction.

How do enzymes work?

Enzyme molecules typically have a cavity in their shape called the **active site**. The reacting molecules (called the **substrate**) fit precisely into this active site and are held in the right orientation to react, see Figure 1. While there, bonds are broken and new ones formed. The reaction takes place with a lower **activation energy** than without the enzyme.

Enzymes are **stereospecific**. They are sensitive to the exact shape of the substrate molecule, and so catalyse the reaction of only one of a pair of optical isomers (**stereoisomers**).

Enzyme inhibition

The shape of an enzyme can easily be altered by heating it up or by changing the pH. These factors disrupt the intermolecular interactions that hold the enzyme in its particular shape and the enzyme loses its catalytic ability – this is called **denaturing**.

Another way to inhibit the activity of an enzyme is to use a molecule of similar shape to the substrate. This fits into the active site and prevents the substrate from entering. Many drugs work like this.

Shapes of complex molecules can be modelled by computers using the principles described in Topic 3.6, The shapes of molecules and ions, and this can be used to design new drugs.

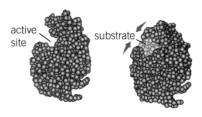

▲ **Figure 1** *The substrate of an enzyme fits its active site*

Synoptic link

Activation energy and its link to rates of reaction are discussed in Topics 5.3, Catalysts and 18.4, The Arrhenius equation.

Synoptic link

Optical isomerism is discussed in Topics 25.2, Optical isomerism and 25.3, Synthesis of optically active compounds.

Common misconception

Increasing pH makes a solution less acidic (more alkaline). *Decreasing* the pH makes a solution more acidic.

Summary questions

1. Suggest how decreasing the pH can disrupt hydrogen bonding between C=O and NH_2 groups shown below.

   $$\text{C}=\text{O}:\text{----H}-\text{N}-$$

   (2 marks)

2. Suggest how increasing the pH can disrupt hydrogen bonding between COOH and NH_2 groups shown below.

   $$-\text{C}(=\text{O})-\text{O}-\text{H}----:\text{N}-$$

   (2 marks)

3. List the following types of bond, which might be found in holding a substrate to an enzyme, in order of decreasing strength (i.e., strongest first).
   Dipole–dipole; hydrogen bond; ionic bond; van der Waals forces.

   (2 marks)

30.4 DNA

Specification reference: 3.3.13

Revision tip
Sometimes it is simpler just to show the basic skeleton of a nucleotide as:

Revision tip
You will not be expected to reproduce the structures of nucleotide bases in exams – they are given in the data sheet.

Synoptic link
This is a condensation polymerisation reaction, see Topic 29.1, Condensation polymers.

What is DNA?

DNA is short for deoxyribonucleic acid, the molecule that is found in all living cells and which carries the 'blueprint' for the organism. Each organism has its own particular of DNA, so there is an infinite variety of DNA molecules.

What are nucleotides?

Each molecule of DNA is a polymer made from just four different monomers, called **nucleotides**. It is the specific order of these monomers that defines each type of DNA.

Nucleotide molecules have three parts, see Figure 1.

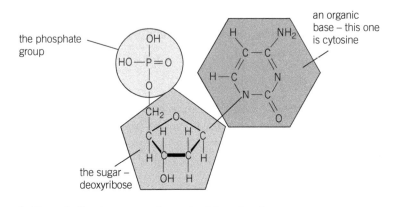

▲ **Figure 1** *The three parts of a nucleotide molecule*

There are four different organic bases:

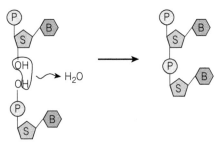

▲ **Figure 2** *Two nucleotides react together to eliminate a molecule of water*

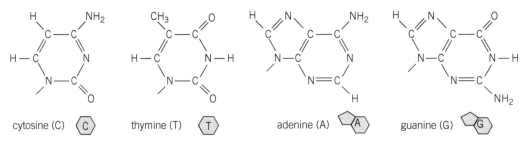

Polymerisation of nucleotides

Two nucleotides react together to eliminate a molecule of water, and further reaction forms a polymer chain – DNA, Figures 2 and 3.

DNA has infinite variations depending on the order of the bases, for example, CCGATTCGTA, etc.

Hydrogen bonding between nucleotides

Adenine and thymine can hydrogen bond together as can guanine and cytosine.

▲ **Figure 3** *Part of a strand of DNA*

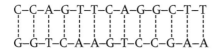

The double helix

Strands of DNA exist in pairs, the strands being held together by hydrogen bonds. The two strands are **complementary**, where one strand has an A the other will have a T, and similarly for G and C.

```
C-C-A-G-T-T-C-A-G-G-C-T-T
| | | | | | | | | | | | |
G-G-T-C-A-A-G-T-C-C-G-A-A
```

Two strands wind together to form a double helix, held together by hydrogen bonds between the bases.

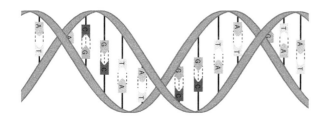

Genetic information is held in DNA coded by the order of the bases. When a cell divides and replicates, the double helix of its DNA unwinds. This is possible because the hydrogen bonds that hold the strands together are much weaker than covalent bonds. Then two new complementary strands of DNA are built round the originals from separate nucleotides found in the cell.

Summary questions

1 State the three component groups of a nucleotide. *(3 marks)*

2 State the complementary sequences of bases to AACCGTGTC. *(1 mark)*

3 What part of a DNA strand is responsible for its acidity? *(1 mark)*

30.5 The action of anti-cancer drugs

Specification reference: 3.3.13

How does cisplatin work?

Cancer cells grow and replicate much faster than normal cells.

Cisplatin is a successful treatment for a number of cancers including testicular cancer. It works by bonding to DNA and disrupting the replication of cells.

Cisplatin is square planar and has the formula

Cisplatin works by bonding to two of the bases in DNA. Nitrogen atoms on two adjacent guanine molecules, of the double helix, act as good ligands and displace the two chloride ligands. This prevents replication of the DNA.

Unfortunately this will also affect the replication of healthy cells, but as cancer cells are replicating faster, the drug has a greater effect on them.

> **Synoptic link**
>
> The shapes of transition metal complexes are discussed in Topic 23.2, Complex formation and the shape of complex ions.

> **Revision tip**
>
> The *cis* in cisplatin tells us that both the chloride ligands (and hence both the ammonia ligands) are on the same side of the complex. It may also be called *Z*-platin using the *E-Z* system, see Topic 14.1, Alkenes.

> **Synoptic link**
>
> Ligand substitution reactions are discussed in Topic 24.2, Ligand substitution reactions.

> **Synoptic link**
>
> Replication is described briefly in Topic 30.4, DNA.

Summary questions

1. What feature must be present on a nitrogen atom for it to act as a ligand? *(1 mark)*

2. What sort of bonds are formed between ligands and metal ions? *(1 mark)*

3. What is the oxidation state of the platinum atom in cisplatin? *(1 mark)*

Chapter 30 Practice questions

1 The partially displayed formula of a tripeptide is shown below.

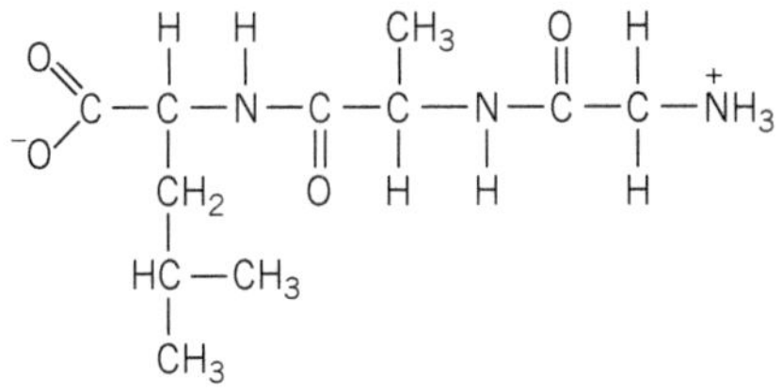

   [structure diagram of tripeptide]

   a What is meant by the term 'displayed formula'? (2 marks)
   b Circle the two peptide (amide) linkages. (2 marks)
   c The left hand amino acid in the structure is leucine, leu. Name the other two (from left to right) and give their three letter abbreviations. (4 marks)
   d The peptide is shown with both a positive and a negative charge. What is the name of this form of a peptide? (1 mark)
   e i How could the tripeptide be split into its constituent amino acids?
      ii What molecule is added to each peptide link during this process? (4 marks)
   f What technique can be used to separate mixtures of amino acids and identify them? (1 mark)
   g What would happen to the structure of the tripeptide in i a strong acid ii a strong base? (2 marks)

2 State three types of bond that help to maintain the tertiary structure of a protein. (3 marks)

3 How many hydrogen bonds form between the bases a guanine (G) and cytosine (C) b adenine (A) and thymine (T). (2 marks)

4 The structure of the amino acid glycine is shown.

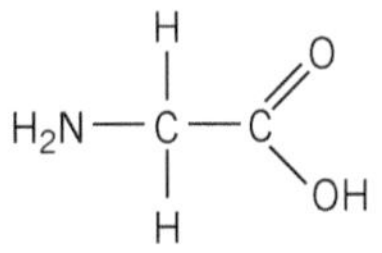

   [structure of glycine]

   a Draw the structure of glycine in i a concentrated acid solution ii a concentrated alkaline solution. (2 marks)
   b Solid glycine has a much higher melting point than would be expected from its relative molecular mass. Explain this observation. (1 mark)
   c Glycine is the only α-amino acid that is not optically active. Explain this observation. (1 mark)
   d State the IUPAC name of glycine. (1 mark)

31.1 Synthetic routes
Specification reference: 3.3.14

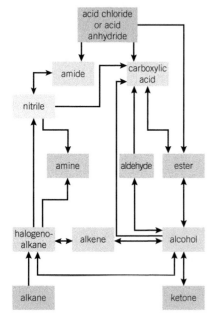

▲ **Figure 1** *Use this chart to revise organic reactions*

Synthesising molecules is about starting with one chemical and by a series of reactions ending up with another – the target molecule.

reaction scheme
Starting material ⟶ target molecule.

Revision tip
Syntheses are done in many chemical industries with the aim of making the process as environmentally clean and also efficient as possible so chemists aim to design processes that:
- do not require a solvent
- use non-hazardous starting materials
- use as few steps as possible
- have a high percentage atom economy.

Working out a scheme
Figure 1 shows how the functional groups are related. Learn the reagents and conditions for each step.

Question and model answer: How can propylamine be synthesised from ethene?

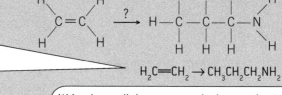

$H_2C=CH_2 \rightarrow CH_3CH_2CH_2NH_2$

> Propylamine has one more carbon atom than ethene. This suggests that the formation of a nitrile is needed at some stage. See Topic 13.2, Nucleophilic substitution in halogenoalkanes.

> Write down all the compounds that can be made from ethene and all the compounds from which the propylamine can be made.

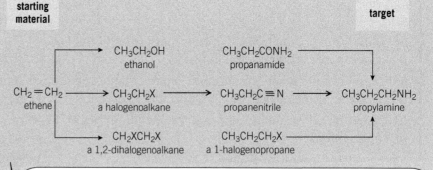

> None of the compounds made in one step from the starting material can then be converted into the product, so you need more than two steps. The formation of a nitrile is required. A halogenoethane can be converted into propanenitrile so you can complete the synthesis in three steps:

Step 1: $CH_2=CH_2 \xrightarrow{HCl} CH_3CH_2Cl$
ethene   chloroethane

Step 2: $CH_3CH_2Cl \xrightarrow{KCN/dil. H_2SO_4} CH_3CH_2C\equiv N$
chloroethane   propanenitrile

Step 3: $CH_3CH_2C\equiv N \xrightarrow{Ni/H_2} CH_3CH_2CH_2NH_2$
propanenitrile   propylamine

Summary questions

1. Why do chemists design syntheses **a** with as few steps as possible **b** with high atom economies? *(2 marks)*

2. Give a two-step scheme to convert 1-chlorobutane to butananoic acid. *(2 marks)*

3. Give a three-step scheme to convert ethane to ethanal. *(3 marks)*

31.2 Organic analysis
Specification reference: 3.3.6

The first step to identifying an organic compound chemically is to find out its functional groups.

Chemical reactions

🧪 Some tests are very straightforward:
- Is the compound acidic (suggests carboxylic acid)?
- Is the compound solid (suggests long carbon chain or ionic bonding), liquid (suggests medium length carbon chain or polar or hydrogen bonding), or gas (suggests short carbon chain, little or no polarity)?
- Does the compound dissolve in water (suggests polar groups) or not (suggests no polar groups)?
- Does the compound burn with a smoky flame (suggests high C : H ratio, possibly aromatic) or non-smoky flame (suggests low C : H ratio, probably non-aromatic)?

Some specific chemical tests are listed in Table 1.

> **Revision tip**
> You will need to know *all* the organic chemistry studied in your A Level course.

> **Synoptic link**
> Look back at Topic 10.3, Reactions of halide ions, for more detail on how the silver halide precipitate, AgX, can be used to identify the halogen.

> **Revision tip**
> You cannot use this test to identify a fluoroalkane because silver(I) fluoride, AgF, is soluble in water.

▼ **Table 1** *Chemical tests for functional groups*

Functional group	Test	Result
alkene −C=C−	shake with bromine water	red-brown colour disappears
halogenoalkane R—X	1. add NaOH(aq) and warm 2. acidify with HNO_3 3. add $AgNO_3$(aq)	precipitate of AgX
alcohol R—OH	add acidified $K_2Cr_2O_7$	orange colour turns green with primary or secondary alcohols (also with aldehydes)
aldehyde R—CHO	warm with Fehling's solution or warm with Tollens' reagent or add acidified $K_2Cr_2O_7$	blue colour turns to red precipitate silver mirror forms orange colour turns green
carboxylic acid R—COOH	add $NaHCO_3$(aq)	bubbles observed as carbon dioxide given off

Summary questions

1. A solid compound burns with a smoky flame and fizzes when a solution of sodium hydrogencarbonate is added.
   a What two functional groups does this suggest? *(2 marks)*
   b The compound's relative molecular mass is found to be 122 to the nearest whole number.
      i What technique could be used to find M_r? *(1 mark)*
      ii What is the likely formula (and name) of the compound? *(2 marks)*
   c If the M_r had been 136, suggest three isomeric compounds that it could be. *(3 marks)*
   d i Suggest another isomer of M_r 136 that does not fizz when sodium hydrogencarbonate is added. *(1 mark)*
      ii What functional group does this compound have? *(1 mark)*
   e A student suggested that the three compounds in **c** could be told apart by high resolution mass spectrometry to determine M_r to four decimal places. Explain why this would not work. *(2 marks)*

Chapter 31 Practice questions

1. The following conversions go *via* an intermediate compound. State a suitable compound in each case.

    a Ethene to ethanoic acid.

    b Ethanol to propanenitrile.

    c Propanone to 2-bromopropane.

    d Benzene to phenylamine. *(4 marks)*

2. State the reactants and conditions for **1a** and **1d** above. *(4 marks)*

3. How would you convert propan-1-ol to propanone (three steps)? *(3 marks)*

4. Give reagents and conditions (including catalyst if appropriate) for each of the steps in the 2-step conversion of 1-chlorobutane to butanoic acid referred to in Q1. *(7 marks)*

5. a In the conversion of 1-chloropropane to butanoic acid, what step is used to increase the carbon chain length by one? *(1 mark)*

    b Name the intermediate compound in this conversion and give its structural formula. *(2 marks)*

    c The intermediate compound can also be converted to 1-butylamine.

       i What reagent and catalyst can be used to bring about this conversion? *(2 marks)*

       ii Why is this a better method of making 1-butylamine than reacting 1-chlorobutane with ammonia? *(2 marks)*

6. The compound $H_2C=CHCH_2COOH$ has two functional groups.

    a Name both functional groups. *(2 marks)*

    b Describe a test (and its result) to confirm the presence of each functional group. *(4 marks)*

    c Give the structural formula of the organic compound formed as a result of each test. *(2 marks)*

32.1 Nuclear magnetic resonance (NMR) spectroscopy

Specification reference: 3.3.15

Nuclear magnetic resonance spectroscopy (NMR) is used mainly in organic chemistry to aid structure determination.

The principles of NMR

Some nuclei, such as ^{1}H and ^{13}C have a property called 'spin'. If they are placed in a magnetic field and subjected to a radio frequency, see Figure 1, the nuclei 'resonate'. The frequency of the resonance depends on the electron density around the atom, which shields the nucleus from the magnetic field. So the resonant frequency depends on the environment of the H or C atom.

An NMR spectrum shows the resonant frequency as a quantity called **chemical shift**, δ, expressed in units called parts per million (ppm) relative to a standard compound called tetramethylsilane, $Si(CH_4)_4$ (TMS). Using Tables of δ values, you can identify how many different types of carbon (or hydrogen) atoms there are in the molecule and what type they are.

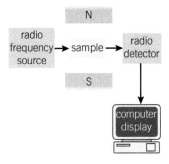

▲ **Figure 1** Schematic diagram of an NMR spectrometer

^{13}C NMR

Only 1% of carbon atoms are carbon-13, but their resonances can be detected and therefore their chemical shifts can be measured. In ^{13}C NMR values of δ range from 0 to around 200 ppm.

^{13}C NMR spectra

Figure 2 shows the ^{13}C NMR spectrum of ethanol. It has two peaks, one for each carbon, because the carbon atoms are in different environments. The one bonded to the oxygen atom has a greater δ value, because the oxygen atom attracts electrons away from this carbon. This carbon atom is deshielded, and has a *greater* δ value than the other carbon. The other carbon, CH_3CH_2OH, is shielded by more electrons and has a *smaller* δ value. Relate the δ values of these peaks to Table C in the data sheet.

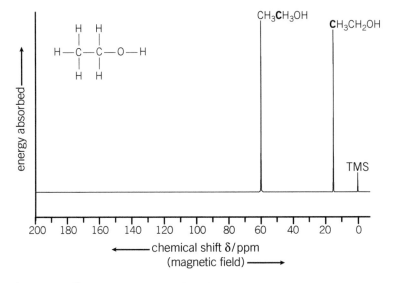

▲ **Figure 2** ^{13}C NMR spectrum of ethanol

Revision tip
The heights of the peaks in ^{13}C NMR spectra are not significant, it is their δ values that are important in interpreting spectra.

Structure determination

Summary questions

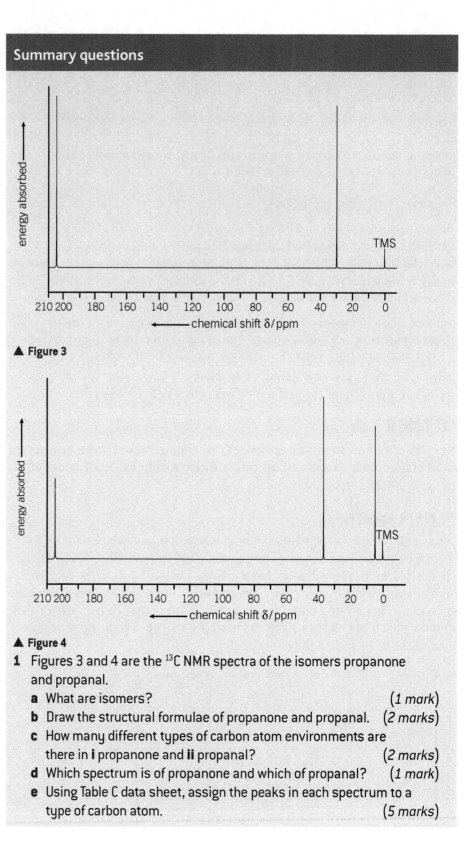

▲ Figure 3

▲ Figure 4

1 Figures 3 and 4 are the ^{13}C NMR spectra of the isomers propanone and propanal.
   a What are isomers? (1 mark)
   b Draw the structural formulae of propanone and propanal. (2 marks)
   c How many different types of carbon atom environments are there in **i** propanone and **ii** propanal? (2 marks)
   d Which spectrum is of propanone and which of propanal? (1 mark)
   e Using Table C data sheet, assign the peaks in each spectrum to a type of carbon atom. (5 marks)

32.2 Proton NMR

Specification reference: 3.3.15

In proton NMR, it is the ^{1}H nucleus that is being examined. It is the hydrogen atoms attached to different functional groups that are resonating.

The ^{1}H NMR spectrum

If all the hydrogen nuclei in an organic compound are in identical environments, you get only one chemical shift value, for example, all the hydrogen atoms in methane, CH_4. But, in a molecule like methanol, CH_3OH, there are hydrogen atoms in two different environments – the three on the carbon atom, and the one on the oxygen atom. The NMR spectrum will show these two environments (see Figure 1).

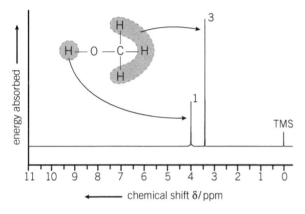

▲ **Figure 1** *The NMR spectrum of methanol – the peak areas are in the ratio 1 : 3*

In ^{1}H NMR the *areas* under the peaks (shown here by the numbers next to them) are proportional to the number of hydrogen atoms of each type – in this case three and one.

The integration trace

The ratio of the peak areas is produced accurately by the NMR spectrometer with a line called **the integration trace**, shown in Figure 2. The relative areas of the peaks of this trace give the relative number of each type of hydrogen, 3 : 1 in this case.

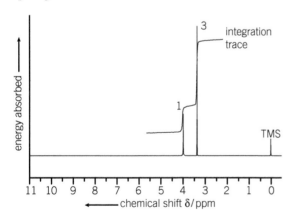

▲ **Figure 2** *The ^{1}H NMR spectrum of methanol showing the integration trace*

The chemical shift value at which the peak representing each type of proton appears tells you about its environment – the type of functional group of which it is a part, see Table B on the data sheet.

Revision tip
^{1}H NMR spectra are run in a solution with either a solvent of tetrachloromethane or one that contains deuterium instead of hydrogen.

Revision tip
The further away a hydrogen atom is from an electronegative atom (such as oxygen) the *smaller* its chemical shift, δ.

Revision tip
The chemical shift values for proton NMR are in the data sheet in exams.

Summary questions

1. Explain why TMS gives only one line in its ^{1}H NMR spectrum. *(1 mark)*

2. **a** How many peaks would you expect to see in the ^{1}H NMR spectrum of 1,4-dimethylbenzene?
   b What would be the ratio of their peak areas?
   c Explain your answers. *(3 marks)*

1,4-dimethylbenzene

32.3 Interpreting proton, ^{1}H, NMR spectra

Specification reference: 3.3.15

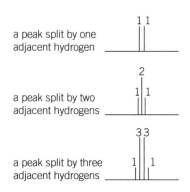

▲ Figure 1 ^{1}H NMR splitting patterns

> **Revision tip**
> Spin–spin coupling is not seen when equivalent hydrogens are on adjacent carbon atoms, for example, $HOCH_2CH_2OH$ has only two single peaks in its ^{1}H NMR spectrum.

Spin–spin coupling

Most ^{1}H NMR peaks are split into particular patterns – called **spin–spin coupling** (also called spin–spin splitting). It happens because the magnetic field felt by any hydrogen atom is affected by the magnetic field of all the hydrogens on the adjacent carbon atoms. So, spin–spin splitting also gives information about the number of neighbouring hydrogen atoms, which helps you to work out structure.

The $n + 1$ rule

Figure 1 shows the spin–spin splitting patterns.

For a particular hydrogen:

- One hydrogen atom on an adjacent carbon will split the NMR signal of a particular hydrogen into two peaks each of the same height.
- Two hydrogen atoms on an adjacent carbon, will split the NMR signal into three peaks with height ratio 1 : 2 : 1.
- Three adjacent hydrogen atoms will split the NMR signal into four peaks with height ratio 1 : 3 : 3 : 1.

This is called the $n + 1$ rule:

n hydrogens on an adjacent carbon atom will split a peak into $n + 1$ smaller peaks.

Interpreting an ^{1}H NMR spectra

Each chemical environment corresponds to a chemical shift value – see Table 1 in Topic 32.2, Proton NMR.

> **Worked example: Analysing a ^{1}H spectrum**
> Look at the peaks in the spectrum in Figure 2.
>
>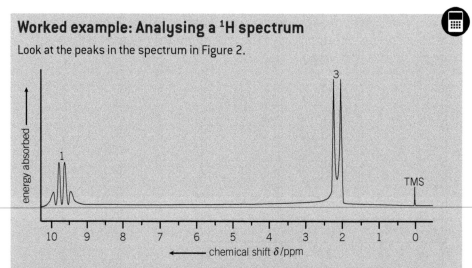
>
> ▲ Figure 2 The ^{1}H NMR spectrum of an organic compound
>
> There are two types of hydrogen environments. See Table B in the data sheet. There is:
>
> - A peak at around δ 9.7. This is the hydrogen of a –CHO group. This peak is split into four (height ratios 1 : 3 : 3 : 1) so there must be an adjacent –CH_3 group.
> - The peak with δ 2.2 is caused by three hydrogens of a –CH_3 group. This peak is split into two (height ratios 1 : 1) by the one hydrogen of the adjacent –CHO group.
>
> So this is ethanal CH_3CHO.

Summary questions

1. The ^{1}H NMR spectrum of ethanol, CH_3CH_2OH, is shown.

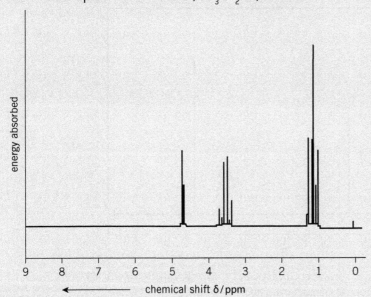

   a. What is the small peak at $\delta = 0$? What is it used for? *(2 marks)*
   b. Use Table B in the data sheet to state which peak belongs to which set of hydrogen atoms. *(3 marks)*
   c. How does the splitting of the peaks confirm your answer to **b**? *(3 marks)*
   d. Explain why the ^{1}H NMR spectrum of ethanol has three peaks while the ^{13}C spectrum has only two. *(2 marks)*

2. Explain why the ^{1}H NMR spectrum of methanol, Figure 1, Topic 32.2, shows no spin–spin coupling. *(1 mark)*

3.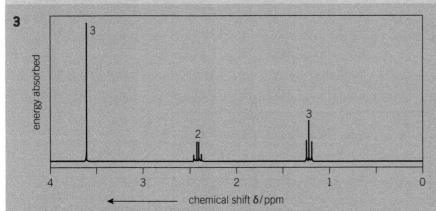

   The ^{1}H NMR spectrum of methyl propanoate, is shown. Using Table B in the data sheet and the spin–spin splitting, assign the peaks in the spectrum to the appropriate groups in the compound. *(6 marks)*

Chapter 32 Practice questions

1 The ¹H NMR spectrum of ethanoic acid, CH_3COOH, is shown below.

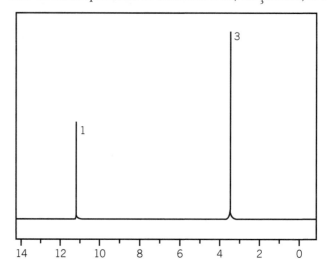

a How many different types of hydrogen atom are present in a molecule of ethanoic acid? *(1 mark)*

b Explain why there is no spin–spin splitting of the two peaks. *(1 mark)*

c What peak is normally found on NMR spectra at δ = 0. What is the purpose of this peak? *(2 marks)*

d Suggest a solvent for running the spectrum of ethanoic acid and explain your choice. *(2 marks)*

2 a How many peaks would you expect in the ¹³C NMR spectrum of pentan-3-one? Explain your answer. *(2 marks)*

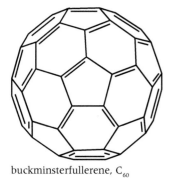

pentan-3-one

b How many peaks would you expect in the ¹H NMR spectrum of pentan-3-one? Explain your answer. *(2 marks)*

c Predict the ratio of peak heights in the ¹H NMR spectrum of pentan-3-one. *(1 mark)*

d Predict how the peaks in the ¹H NMR spectrum of pentan-3-one will be split. Explain your answer. *(4 marks)*

e Name the rule that predicts spin–spin splitting of ¹H NMR peaks. *(1 mark)*

3 The ¹³C NMR spectrum of the compound buckminsterfullerene, C_{60}, is shown.

buckminsterfullerene, C_{60}

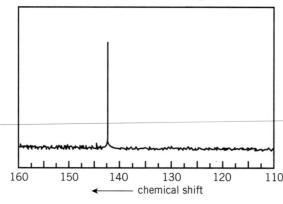

a The spectrum has one line only. What does this confirm about the environment of each carbon atom in C_{60}? *(1 mark)*

b What compound is used in NMR spectroscopy to calibrate the zero of the spectrum? State its name (or abbreviation) and its formula. *(2 marks)*

c What are the units of chemical shift used in the spectrum? *(1 mark)*

d How many peaks would you expect to see in the ¹H NMR spectrum of buckminsterfullerene? Explain your answer. *(2 marks)*

33.1 Chromatography
Specification reference: 3.3.16

Chromatography is a family of techniques for separating and identifying the components in a mixture. Types of chromatography include:

- Thin-layer chromatography (TLC) – a plate is coated with a solid and a solvent moves up the plate.
- Column chromatography (CC) – a column is packed with a solid and a solvent moves down the column.
- Gas chromatography (GC) – a column is packed with a solid or with a solid coated by a liquid, and a gas is passed through the column.

All these techniques have in common:

- The **mobile phase** which carries the soluble components of the mixture with it. The more soluble the component in the mobile phase, the faster it moves.
- The **stationary phase** which holds back the components in the mixture that are attracted to it. The more affinity the component has for the stationary phase, the slower it moves.

So, if suitable moving and stationary phases are chosen, a mixture of similar substances can be separated completely, because every component of the mixture will move over the solid at a different rate.

Thin-layer chromatography

The mobile phase is a solvent. The stationary phase is a thin layer of silica gel (silicon dioxide, SiO_2) or alumina (aluminium oxide, Al_2O_3) on a glass, metal, or plastic sheet (see Figure 1).

Synoptic link
TLC is often used for separating and identifying mixtures of amino acids see Topic 30.2, Peptides, polypeptides, and proteins.

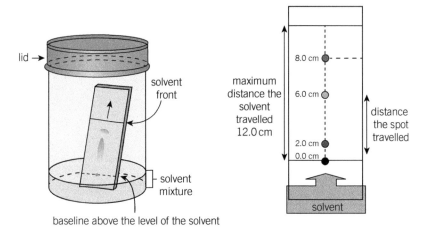

▲ **Figure 1** *A TLC experiment and the resulting chromatogram*

Revision tip
All R_f values must be less than 1.

When the chromatogram has run, the position of colourless spots may have to be located by shining ultraviolet light on the plate, or chemically by spraying the plate with a locating agent which reacts with the components of the mixture to give coloured compounds.

After the plate has been run, an R_f value is calculated of each component using:

R_f = distance moved by spot / distance moved by solvent

The R_f values can be used to help identify each component.

Chromatography

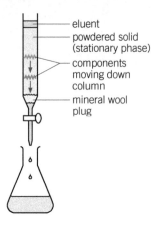

▲ Figure 2 *Column chromatography*

Column chromatography

The stationary phase is a powder, such as silica, aluminium oxide, or a resin, packed into a narrow tube – the column. A solvent (the eluent) is added at the top (see Figure 2). As the eluent runs down the column, the components of the mixture move at different rates and collect separately in flasks at the bottom. A mixture of amino acids can be separated into its pure components by this method.

Gas–liquid chromatography (GC)

The basic apparatus is shown in Figure 3.

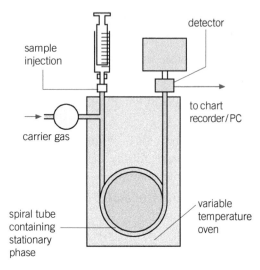

▲ Figure 3 *Gas–liquid chromatography (GC)*

Synoptic link

The analytical use of mass spectrometry was covered in Topic 16.2, Mass spectrometry.

The stationary phase is a powder, coated with oil and packed into (or coated onto) the inside of a long coiled capillary tube. This is placed in an oven. The mobile phase is a gas, such as nitrogen or helium which carries the mixture. This separates the components as some move along with the gas and some are retained by the oil. They are detected as they leave the column at different times – called their **retention times**. The results are presented on a graph (Figure 4) – the area under each peak is proportional to the amount of that component.

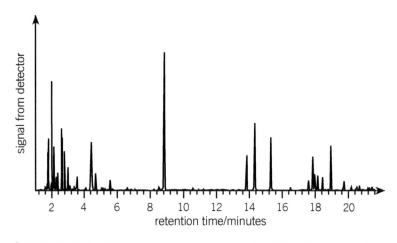

▲ Figure 4 *Typical GC trace – each peak represents a different component*

Summary questions

1 Calculate the R_f values of each of the spots in Figure 1. *(3 marks)*

2 Why must all R_f values be less than 1? *(1 mark)*

3 Why is chromatography usually the best method for separating amino acids? *(1 mark)*

GCMS — Gas chromatography–mass spectrometry

If a mass spectrometer is used as the detector, each component is fed automatically into the mass spectrometer and the compound is identified either by its fragmentation pattern or by measuring its accurate mass.

Chapter 33 Practice questions

1 A thin-layer chromatogram of three separate amino acids is shown.

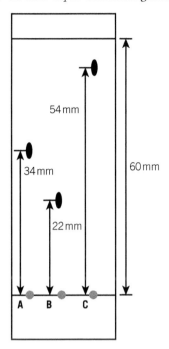

   a i Define R_f value.
      ii Calculate the R_f value for each amino acid. *(4 marks)*
   b i Using the table of R_f values below, suggest which amino acid (A, B, or C) is most likely to be threonine.

Amino acid	Rf value
alanine	0.21
arginine	0.21
threonine	0.34
tyrosine	0.43

      ii What must be kept the same for your conclusion to be valid? *(2 marks)*
   c Amino acids are colourless. State two ways by which the spots can be located. *(2 marks)*
   d Explain why all R_f values must be less than 1. *(1 mark)*

2 Complete the table about different types of chromatography. *(4 marks)*

	Stationary phase	Mobile phase
Thin layer		Liquid solvent
Column	Silica or alumina or resin	
GLC		

3 Describe how you would run a TLC experiment to separate a mixture of amino acids. *(6 marks)*

Synoptic questions

1. In the 1930s, a CFC gas, CCl_2F_2, was introduced by the American chemist Thomas Midgley as a safe replacement for refrigerant gases such as ammonia and sulfur dioxide. He inhaled a lungful of the gas and used it to blow out a candle to demonstrate that it is non-toxic and non-flammable.

   a What does CFC stand for? *(1 mark)*

   b CFCs were originally used as refrigerants. State two other uses to which CFCs have been put. *(2 marks)*

   c CFCs in the atmosphere are responsible for the breakdown of ozone in the upper atmosphere. What environmental problems does this cause? *(2 marks)*

   d In the stratosphere CFCs break down to give chlorine atoms which catalyse the following reactions:

   $Cl\bullet (g) + O_3(g) \rightarrow ClO\bullet (g) + O_2(g)$ reaction 1

   $ClO\bullet (g) + O(g) \rightarrow O_2(g) + Cl\bullet (g)$ reaction 2

   i What is the significance of the dot on the species $ClO\bullet$ and $Cl\bullet$? *(1 mark)*

   ii What name is given to such species? *(1 mark)*

   iii Write the overall equation resulting from reactions 1 and 2. *(2 marks)*

   iv State the function of $Cl\bullet$ in the reactions. Explain your answer. *(2 marks)*

   e The above two reactions are the **propagation** steps of a chain reaction.

   i Write an equation for a **termination** step that removes $Cl\bullet$ from the atmosphere. *(1 mark)*

   ii Suggest why this termination step is slow. *(1 mark)*

   f Reaction 1 has been found to be first order with respect to both $Cl\bullet$ and O_3.

   i Write an expression for the rate equation for this reaction. *(1 mark)*

   ii What is the overall order of this reaction? *(1 mark)*

   g Since the Montreal Protocol of 1987, CFC production has been phased out. Suggest why there is still a reservoir of CFCs in the atmosphere. *(2 marks)*

   h CF_2I_2 may be a possible replacement for CCl_2F_2. Explain why CF_2I_2 has a higher boiling point than CCl_2F_2. *(2 marks)*

   i When CF_2I_2 reacts with sodium hydroxide solution, a C–I bond is more likely to break than a C–F bond. Explain why you would expect this. *(1 mark)*

   j Describe a test for I^- ions. Explain what must be done to the reaction mixture before this test can be carried out. *(5 marks)*

2. Compound A is the halogenoalkane $CH_3CH_2CHBrCH_3$, B is the alkene $CH_3CH_2CH=CH_2$, and C is the alcohol $CH_3CH_2CHOHCH_3$.

   a State the IUPAC name of these compounds. *(3 marks)*

   b State the reactions and conditions required to convert a halogenoalkane into **i** an alkene **ii** an alcohol. State the type of reaction in each case. *(8 marks)*

c When A is converted into B, a second alkene is also formed which is an isomer of B.

    i Give the structural formula and IUPAC name of this second alkene. *(1 mark)*

    ii What sort of isomerism is shown by this alkene which is not shown by B? *(1 mark)*

    iii Draw the displayed formulae of the two isomers it can form and name them. *(4 marks)*

d Compound A can be converted into compound D the carboxylic acid $CH_3CH_2CH(COOH)CH_3$, in two steps. Outline the two steps. *(2 marks)*

e **i** Draw the displayed formula of $CH_3CH_2CH(COOH)CH_3$.

    ii This compound shows optical isomerism. Explain this, marking a star on the chiral carbon atom. *(3 marks)*

3 In a petrol-driven car, a mixture of hydrocarbon fuel and air is ignited by a spark. The exhaust fumes contain the pollutant gases nitrogen monoxide, NO, nitrogen dioxide, NO_2, carbon monoxide, CO, and unburnt hydrocarbons. The amount of these pollutants can be reduced using a catalytic converter consisting of a ceramic mesh coated with the catalyst – a mixture of platinum and rhodium.

  a What environmental problems do these gases cause? *(3 marks)*

  b Explain how the exhaust gases contain carbon monoxide. *(1 mark)*

  c Write balanced symbol equations with state symbols for the formation of **i** nitrogen monoxide and **ii** nitrogen dioxide. *(4 marks)*

  d Nitrogen and carbon dioxide leave the catalytic converter. Explain why they are less polluting. *(2 marks)*

  e What is the oxidation state of the nitrogen atom in **i** N_2 **ii** NO **iii** NO_2? *(3 marks)*

  f Suggest why nitrogen gas, N_2, will only react at high temperatures. *(2 marks)*

  g Is the catalyst in the converter a heterogeneous or homogeneous catalyst? Explain these terms *(3 marks)*

  h State in general terms how a catalyst speeds up a reaction. *(1 mark)*

  i Explain why the catalyst is coated onto a ceramic mesh. *(1 mark)*

  j One component of the petrol fraction is the octane 2,2,4-trimethylpentane. Draw the skeletal formula for this hydrocarbon. *(1 mark)*

  k In the past, petrol had the compound tetraethyl lead added to it to promote the even burning of the fuel. Explain why this is no longer used. *(1 mark)*

  l In cold weather, a colourless liquid can be seen dripping from the exhaust pipes of cars just after they have started. What is this liquid and how is it formed? *(2 marks)*

4 Lithium hydroxide is used in spacecraft to remove exhaled carbon dioxide from the air by the following reaction:

$2LiOH(s) + CO_2(g) \rightarrow Li_2CO_3(s) + H_2O(l)$

  a Sodium hydroxide undergoes a similar reaction. Explain why lithium hydroxide is preferred in spacecraft. *(2 marks)*

Synoptic questions

b Write the equation for sodium hydroxide absorbing carbon dioxide. *(1 mark)*

c Typically a human being exhales around 500 dm³ of carbon dioxide per day.

  i How many moles of CO_2 is this? (1 mol of any gas has an approximate volume of 24 dm³ under room conditions.) *(1 mark)*

  ii How many moles of LiOH are needed to absorb this amount of CO_2? *(1 mark)*

  iii How many grams of LiOH is this? *(1 mark)*

  iv How many grams of sodium hydroxide would be needed to absorb the same amount of carbon dioxide? *(1 mark)*

d In the laboratory, calcium oxide is often used to absorb carbon dioxide.
$$CaO(s) + CO_2(g) \rightarrow CaCO_3(s)$$

The table gives values of enthalpies of formation and entropies of the compounds involved.

	CaO(s)	CO_2(g)	$CaCO_3$(s)
$\Delta_f H^{\ominus}$ / kJ mol⁻¹	−635.1	−393.5	−1206.9
$S^{\ominus}$ / J K⁻¹ mol⁻¹	39.7	213.6	92.9

  i Explain why CO_2(g) has the largest value of entropy. *(1 mark)*

  ii Calculate $\Delta S^{\ominus}$ for the reaction. Explain the sign of your answer. *(2 marks)*

  iii Calculate $\Delta H^{\ominus}$ for the reaction. *(1 mark)*

  iv Calculate $\Delta G^{\ominus}$ at room temperature, 298 K. *(2 marks)*

  v Is the reaction feasible at room temperature? Explain your answer. *(2 marks)*

  vi Calculate $\Delta G^{\ominus}$ at 1500 K. Is the reaction feasible at this temperature? *(3 marks)*

  vii What does the value of ΔG tell you about the reaction rate? *(1 mark)*

5 The anti-inflammatory medicine ibuprofen has the formula:

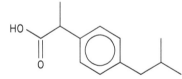

The compound is said to show **optical isomerism**. It has a **chiral centre** and is produced and sold as a **racemate**.

a Explain each of the terms in bold. *(6 marks)*

b Mark the chiral centre with a *.

c How may two optical isomers be distinguished? *(1 mark)*

d Only one isomer of ibuprofen is an active drug. Suggest why ibuprofen is sold as a racemate. *(2 marks)*

e Ibuprofen is not very soluble in water, a fact that slows down its absorption into the bloodstream. Explain why ibuprofen is not very water-soluble. *(1 mark)*

f Ibuprofen can be made more water-soluble (and therefore quicker to act) by reacting it with the α-amino acid lysine which is water-soluble.

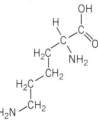

  i What is an α-amino acid? *(2 marks)*
  ii Explain why lysine is water soluble. *(1 mark)*
  iii Why would you expect lysine to be non-toxic? *(1 mark)*

g The two form a compound 'ibuprofen express', which is absorbed more quickly. Suggest how lysine reacts with ibuprofen. (No equation is necessary.) *(1 mark)*

h The M_r values are: ibuprofen, 206; lysine, 146; compound A, the compound formed between lysine and ibuprofen, 334.

  i One tablet of ibuprofen contains 200 mg of ibuprofen. How much ibuprofen express is needed to give the same dose of ibuprofen? *(2 marks)*
  ii The M_r of compound A is 18 less than the sum of the M_rs of lysine and ibuprofen. Suggest an explanation for this. *(1 mark)*

6 4.6 g of an organic compound, A, containing carbon, hydrogen, and oxygen only were burned in excess oxygen in order to determine its formula. The products were 8.8 g carbon dioxide and 5.4 g water.

  a How many moles of **i** carbon dioxide **ii** water, were formed? *(2 mark)*
  b i How many moles of carbon were in the sample of A? *(1 mark)*
  ii How many grams of carbon is this? *(1 mark)*
  c i How many moles of hydrogen were in the sample of A? *(1 mark)*
  ii How many grams of hydrogen is this? *(1 mark)*
  d i What is the total mass of hydrogen and carbon in 4.6 g of A? *(1 mark)*
  ii How many grams of oxygen were in 4.6 g of A? *(1 mark)*
  iii How many moles of oxygen is this? *(1 mark)*
  e What is the empirical formula of A? *(1 mark)*
  f The relative molecular mass of A was found to be 46.0.
    i Name an instrumental technique for determining relative molecular mass. *(1 mark)*
    ii What is the molecular formula of A? *(1 mark)*
  g The two possible structural formulae that A could have are CH_2CH_3OH and CH_3OCH_3. The infrared spectrum of A is shown:

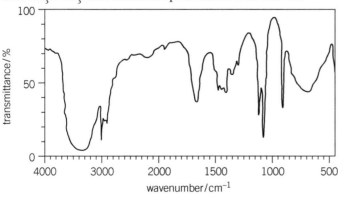

Synoptic questions

      i What is the significance of the broad peak at 3400 cm^{-1}? *(2 marks)*

      ii Which of the two possible formulae is correct? *(1 mark)*

      iii What is the name of A, and to what homologous series does it belong? *(1 mark)*

  h Suggest a chemical test to confirm your answer to **g iii** and give the result you would expect. *(2 marks)*

  i Why would high resolution mass spectrometry (to four decimal places) *not* be able to distinguish between CH_2CH_3OH and CH_3OCH_3? *(2 marks)*

  j How many peaks would you expect to see in the ^1H NMR spectrum of **i** CH_2CH_3OH and **ii** CH_3OCH_3? *(2 marks)*

  k How many peaks would you expect to see in the ^{13}C NMR spectrum of **i** CH_2CH_3OH and **ii** CH_3OCH_3? *(1 mark)*

7 Kleeno is a brand of toilet cleaner sold to remove lime scale from the toilet bowl. Lime scale contains calcium carbonate, $CaCO_3$, which forms a whitish deposit. Kleeno contains hydrochloric acid which dissolves calcium carbonate.

  a Write an equation for the reaction of calcium carbonate with hydrochloric acid. *(2 marks)*

  b Kleeno claims to contain at least 10% hydrochloric acid, i.e. 10 g HCl in 100 cm³ of Kleeno.

      i What is the M_r of hydrochloric acid? *(1 mark)*

      ii How many moles of HCl are there in 10 g of it? *(1 mark)*

      iii What is the concentration of HCl in Kleeno in mol dm^{-3}? *(1 mark)*

  c A consumer organisation decided to test the claim on the bottle that Kleeno contained 10% HCl by doing a titration. They took 25.00 cm³ samples of Kleeno and titrated them with 2.5 mol dm^{-3} sodium hydroxide solution, NaOH, using phenolphthalein as indicator. The average titre was 28.00 cm³.

      i Write the equation for HCl reacting with NaOH. *(1 mark)*

      ii What colour change would take place at the end point? *(1 mark)*

      iii Why is phenolphthalein a better choice than methyl orange for this titration? *(1 mark)*

  d **i** How many moles of NaOH are present in 28.00 cm³ of 2.5 mol dm^{-3} solution? *(1 mark)*

      ii How many moles of HCl are present in 25 cm³ of Kleeno? *(1 mark)*

      iii What is the concentration of HCl in Kleeno? *(1 mark)*

      iv Is the claim on the bottle true? *(1 mark)*

  e The Kleeno bottle warns that it should not be mixed with bleach-based toilet cleaners that contain HClO. This is because the reaction

$$HClO(aq) + HCl(aq) \rightarrow Cl_2(g) + H_2O(l)$$

will occur.

      i State the oxidation states of each atom in the above equation. *(2 marks)*

      ii What problem is caused by this reaction? *(1 mark)*

      iii What approximate volume of chlorine would be produced by adding 50 cm³ of Kleeno to excess bleach? *(2 mark)*

8 This question is about the element magnesium and its compounds. The successive ionisation energies of magnesium are given in the table.

Electron removed	Ionisation energy / kJ mol^{-1}
1st	738
2nd	1451
3rd	7733
4th	10 541
5th	13 629
6th	17 995
7th	21 704
8th	25 657
9th	31 644
10th	35 463
11th	169 996
12th	189 371

a Write an equation to represent the second ionisation energy of magnesium. *(2 marks)*

b Explain the general increase in ionisation energy as electrons are removed from the Mg atom. *(2 marks)*

c Explain how the data in the table support the electron arrangement of magnesium as 2, 8, 2. *(2 marks)*

d The mass spectrum of magnesium is shown below:

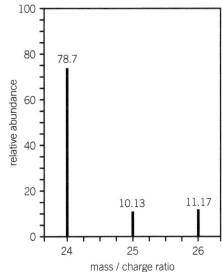

Explain what each line represents. *(2 marks)*

e Calculate the relative atomic mass of magnesium. *(2 marks)*

f State how many protons, neutrons, and electrons there are in an atom of ^{25}Mg. *(3 marks)*

g i Magnesium reacts slowly with water to form magnesium hydroxide. Write a balanced symbol equation for this reaction. *(2 marks)*

  ii Give the oxidation states of each atom in this reaction on each side of the arrow. *(2 marks)*

h Describe the structure and bonding of magnesium oxide. *(2 marks)*

i Magnesium hydroxide is almost insoluble in water. Describe the trend in solubility of the hydroxides of the Group 2 metals as we descend the group. *(1 mark)*

Answers to practice questions

Chapter 1

1 a $1s^2\ 2s^2\ 2p^6\ 3s^2$ [1]   b $1s^2\ 2s^2\ 2p^6$ [1]

2 $\left(\dfrac{78.0}{100} \times 24\right) + \left(\dfrac{10.0}{100} \times 25\right) + \left(\dfrac{11.0}{100} \times 26\right) = 24.34$

24.3 [2]

3

Sub-atomic particle	Relative charge	Relative mass
proton	+1	1	[1]
neutron	0	1	[1]
electron	−1	$\dfrac{1}{1836}$	[1]

4 a Isotopes have the same number of protons/atomic number. But a different number of neutrons/mass number [1]

  b Mass spectroscopy. [1]

  c $\left(\dfrac{25.0}{100} \times 37\right) + \left(\dfrac{75.0}{100} \times 35\right) = 35.5$ [2]

  36

  d Cl [1]

5 Ar [1]   6 D

Chapter 2

1 a $\dfrac{2.00}{55.8} = 0.0358$ mol [1]

  b $0.0358 \times 6.02 \times 10^{23} = 2.16 \times 10^{22}$ [1]

2 a The simplest whole number ratio of the atoms of each element present in the compound. [1]

  b

P	Cl
$\dfrac{14.9}{31.0}$	$\dfrac{85.1}{35.5} = 2.40$
$\dfrac{0.481}{0.481} = 1$	$\dfrac{2.40}{0.481} = 5$

[2]

  PCl_5

3 a $\dfrac{25}{1000} \times 0.1 = 0.0025$ [2]

  b 0.0025 [1]

  c $\dfrac{0.0025 \times 1000}{23.5} = 0.106\,mol\,dm^{-3}$ [2]

4 a Zinc dissolves and bubbles. [2]

  b $\dfrac{1.50}{65.4} = 0.0229$ [2]

  c $\dfrac{50}{1000} \times 0.2 = 0.01$ [2]

  d Yes, the zinc is in excess. [1]

5 B [1]   6 B [1]   7 B [1]

Chapter 3

1 a Tetrahedral.

   It has four bonded pairs of electrons around the central carbon atom.

   It adopts this shape so the pairs of electrons are as far away from each other as possible. [3]

  b 107°

   It has three bonded pairs and one lone pair of electrons around the central nitrogen atom. Lone pairs repel more than bonded pairs. [3]

2 The ability of an atom to attract the electrons in a covalent bond to itself. [2]

3 a Metallic bonds between Mg^{2+} ions and delocalised electrons. [4]

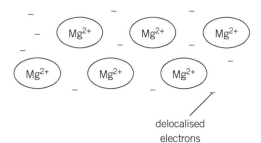

  b The delocalised electrons can move. [1]

4 Water has hydrogen bonds between the lone pair of electrons on the oxygen atom of one molecule and the δ+ H atom of another molecule.

[4]

5 a

Substance	Mg	MgO
structure	giant metallic	giant ionic
type of bonding	metallic	ionic
electrical conductivity when solid	very good	poor

[3]

  b The ions can move. [1]

  c There are strong ionic bonds between the ions. Lots of energy is required to break the bonds between these ions. [2]

6 a The attraction between oppositely charged ions. [2]

  b 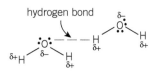 [1]

  c There are lots of strong ionic bonds between the ions, so lots of energy is required to overcome these forces of attraction. [1]

7 B [1]   8 B [1]

Chapter 4

1 a Energy is required to break bonds. [1]

  b energy in = $(4 \times 413) + 612 + 193 = 2457$
     energy out = $(4 \times 413) + (2 \times 285) + 348 = 2570$
     = $-113\,kJmol^{-1}$ [3]

  c It is exothermic. [1]

2 a Bigger surface area / reacts faster. [1]

Answers to practice questions

b The amount of energy needed to raise the temperature of 1 g of the substance by 1 K. [2]

c $100 \times 4.18 \times 11.0$

$= \frac{4598\,J}{4.598\,K}$ or 4.598 kJ [2]

d $\frac{100}{1000} \times 1.00 = 0.100\,mol$ [1]

e $\frac{4.598}{0.1} = -45.98\,kJ\,mol^{-1}$

$= -46.0\,kJ\,mol^{-1}$ [2]

f Heat loss

Add insulation [2]

3 a $C_2H_5OH(l) + 3O_2(g) \rightarrow 2CO_2(g) + 3H_2O(l)$ [2]

b $250 \times 4.18 \times 30$

31350 J / 31.35 kJ or 31.35 kJ [2]

c RMM of ethanol = 46.0

$\frac{2.00}{46.0} = 0.0435$ [2]

d $\frac{31.35}{0.0435} = -721\,kJ\,mol^{-1}$ [2]

4 a The enthalpy change for a reaction is independent of the route taken. [1]

b Lots of different hydrocarbons would be made. [1]

Chapter 5

1 a Energy, E [1]

b The peak must be shifted to the right of the original and must be lower in height than the original. [2]

c The particles gain energy / move faster so they collide more often and when they do collide more of the particles have the activation energy / enough energy to react. [3]

2 a The particles do not have enough energy / the activation energy. [1]

The particles are not in the correct orientation. [1]

b A line drawn to the left of and parallel to the original activation energy, E_a line. [2]

c Catalysts offer an alternative reaction pathway with a lower activation energy. [1]

The diagram shows how now a much greater proportion of the molecules have enough energy to react. [1]

3 a For a reaction to take place the particles have to collide and have enough energy to react/ activation energy. [2]

b Increase the temperature.

Increase the pressure/decrease the volume. [2]

Chapter 6

1 a

	N_2	H_2	NH_3
Amount at start	3.00	6.00	0.00
Amount at equilibrium	2.00	3.00	2.00
Concentration at equilibrium	2.00	3.00	2.00

[1]

b $K_c = \frac{[NH_3(g)]^2}{[N_2(g)][H_2(g)]^3}$ [1]

c $K_c = \frac{2.00^2}{2.00 \times 3.00^3} = 0.0741\,dm^6\,mol^{-2}$ [3]

d Decrease. The forward reaction is exothermic. [1]

e To the right. The right-hand side has fewer gas molecules. [1]

2 a Where nothing can enter or leave. [1]

b It is equal to the rate of backward reaction. [1]

c To the left. It is the endothermic direction. [2]

d No effect. Catalysts increase the rate of forwards and backwards reaction by the same amount. [2]

3 a i Increase. Particles collide more often / more particles have enough energy to react. [1]

ii Moves to left hand-side. This is the endothermic direction. [1]

b i Increase. The particles collide more often. [1]

ii Moves to right-hand side. This side has fewer gas molecules. [1]

c i Increases. More particles have enough energy to react. [1]

ii No effect. Catalysts increase the rate of forwards and backwards reaction by the same amount. [1]

Chapter 7

1 a i 3+ [1] ii 2+ [1]

b i iron(III) chloride [1] ii iron(II) carbonate [1]

2

Substance	Oxidation state of S
H_2S	-2
SO_2	$+4$
SO_3	$+6$
H_2SO_4	$+6$
S_8	0

[5]

3 a 0 [1]

b Iron is oxidised 0 to +2. [1]

Copper is reduced +2 to 0. [1]

c copper sulphate [1]

4 a The metal dissolves. [1]

Bubbles are produced. [1]

b Magnesium loses electrons so is oxidised. [1]

Hydrogen gains electrons so is reduced. [1]

5

	oxidation state
Mg in $MgCO_3$	+2
Cu in Cu_2O	+1
Cl in $NaClO_4$	+7

[3]

6 C [1] 7 B [1]

Answers to practice questions

Chapter 8

1.
Element	Bonding	Structure
Na	metallic	giant metallic lattice
Mg	metallic	giant metallic lattice
Al	metallic	giant metallic lattice
Si	covalent	giant covalent
P_4	covalent	simple molecules
S_8	covalent	simple molecules
Cl_2	covalent	simple molecules
Ar	Van der Waals	monatomic

[2]

2. a Sodium has a giant metallic structure.
   Chlorine has a simple molecular structure.
   The van der Waals' forces between chlorine molecules are weaker than the metallic bonds in sodium. [3]
   b Aluminium has stronger metallic bonds than sodium.
   More energy is required to break the bonds in aluminium than sodium. [3]
3. Sulfur [1]
   Largest molecules / most electrons. [1]
   Strongest van der Waals' forces between molecules. [1]
4. Aluminium
   There is a big jump between the third and fourth ionisation energies so it must be in Group 3. [1]
5. Silicon has a giant covalent structure. [1]
   There are lots of strong covalent bonds between the silicon atoms. [1]
   A lot of energy is needed to overcome these strong forces of attraction. [1]
6. D [1]

Go Further

1. $N(g) \rightarrow N+(g) + e-$
2. The outer electron configuration of oxygen has a pair of electrons in one of the p orbitals. There is repulsion between these electrons so less energy is required to remove an electron.

Chapter 9

1. The first ionisation energy decreases down the group.
   Although the number of protons increases, the outer electrons are further from the nucleus so the atomic radius increases and there is more shielding so the outer electron is lost more easily and the first ionisation energy decreases down the group. [4]
2. Mg is oxidised 0 to +2.
   H is reduced +1 to 0. [2]
3. a Increases. [1] b Decreases. [1] c Decreases. [1]
4. To treat indigestion / heartburn.
   It neutralises excess stomach acid. [2]
5. $Sr^+(g) \rightarrow Sr^{2+}(g) + e^-$
   [1 mark for species]
   [1 mark for state symbols]
6. a $1s^2\ 2s^2\ 2p^6\ 3s^2$ [1]
   b $1s^2\ 2s^2\ 2p^6\ 3s^2\ 3p^6\ 4s^2$ [1]
   c $1s^2\ 2s^2\ 2p^6\ 3s^2\ 3p^6$ [1]
7. B [1] 8. B [1]

Chapter 10

1. Add nitric acid. [1]
   Then add silver nitrate. [1]
   A white precipitate of AgCl means the halide was a chloride. [1]
   A cream precipitate of AgBr means the halide was a bromide. [1]
   A yellow precipitate of AgI means the halide was a iodide. [1]
2. a Increases. [1] b Increases. [1] c Increases. [1]
3. a Red / brown colour appears. [1]
   b Cl is reduced 0 to −1. [1]
   Br is oxidised −1 to 0. [1]
   c chlorine [1]
4. D [1] 5. C [1] 6. D [1]

Chapter 11

1. a A compound that contains hydrogen and carbon only. [1]
   b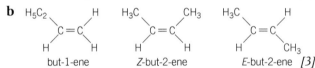
   but-1-ene   Z-but-2-ene   E-but-2-ene [3]
2. 2-bromopropane   1-bromopropane [2]
3. C [1]
4. a They have the same functional group.
   Each successive member differs by the addition of CH_2. [2]
   b C_nH_{2n+2} [1] c $C_{10}H_{22}$ [2]
5. a The simplest whole number ratio of the atoms of each element in the compound. [1]
   b
C	H
$\frac{83.7}{12.0} = 6.975$	$\frac{16.3}{1.0} = 16.3$
$\frac{6.975}{6.975} = 1$	$\frac{16.3}{6.975} = 2.34$
C_3H_7	

[3]
   c C_6H_{14} [1]
6. C [1]
7. a Structural isomers have the same molecular formula but a different structural formula. [1]
   b propan-2-ol [1]
8. D [1]

Answers to practice questions

Chapter 12

1. **a** A compound that contains hydrogen and carbon only and only contains single bonds. *[2]*
   b pentane *[1]*
2. **a** A substance that can be burnt to release energy. *[1]*
   b $CH_4(g) + 2O_2(g) \rightarrow CO_2(g) + 2H_2O(l)$
   [products = 1 mark]
   [balancing = 1 mark]
   c If the fuel contains sulfur when it is burnt sulfur dioxide is formed. *[1]*
   This gas reacts with water to produce acid rain. *[1]*
3. **a** zeolite / silicon dioxide and aluminium oxide *[1]*
   b It has a large surface area. *[1]*
4. As the carbon chain length increases the boiling point increases. *[1]*
   The molecules get larger / have more electrons. *[1]*
   The van der Waals forces are stronger. *[1]*
5. **a** $C_3H_8(g) + 5O_2(g) \rightarrow 3CO_2(g) + 4H_2O(g)$ *[2]*
   b soot / carbon / carbon monoxide *[1]*
   c Insufficient supply of oxygen. *[1]*
6. **a** $CH_4 + Cl_2 \rightarrow HCl + CH_3Cl$ *[1]*
   b A species with an unpaired electron. *[1]*
   c UV light *[1]*
   d $CH_3\bullet + Cl_2 \rightarrow CH_3Cl + Cl\bullet$ *[1]*
   e $2CH_3\bullet \rightarrow C_2H_6$ *[1]*
7. B *[1]*   **8** A *[1]*

Chapter 13

1.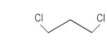
   1,2-dichloropropane   1,3-dichloropropane   2,2-dichloropropane

   1,1-dichloropropane *[4]*

2. Warm the sample with sodium hydroxide. *[1]*
   Add nitric acid. *[1]*
   Then add silver nitrate. *[1]*
   A cream precipitate of silver bromide shows the halogenoalkane was a bromoalkane. *[1]*
3. **a** 1-chloroethane *[1]*
   1-chlorobutane *[1]*
   b 1-chlorobutane *[1]*
   Larger molecule / more electrons. *[1]*
   Stronger van der Waals'. *[1]*
4. **a** A species with a lone pair of electrons that it can donate to form a new covalent bond. *[1]*

   b (mechanism diagram)

   [One mark for dipoles.]
   [One mark for curly arrow from lone pair of oxygen to δ+ C.]
   [One mark for curly arrow breaking the C–Br bond.]
   [One mark for the products.]
   c Nucleophilic substitution. *[2]*

5. **a** (mechanism diagram)

   [One mark for arrow from lone pair on oxygen to H.]
   [One arrow for curly arrow breaking C–H bond.]
   [One mark for curly arrow breaking C–Cl bond.]
   [One mark for products.]
   b but-1-ene *[1]*
   c elimination *[4]*
6. A *[1]*   **7** A *[1]*

Chapter 14

1. **a** chloroethane *[1]*
   b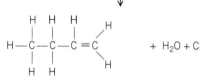

   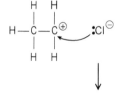

   [5]

Answers to practice questions

2 a [structure of poly(phenylethene) repeat unit] [1]

b Addition polymerisation. [1]

3 a A substance that can be burnt to release energy. [1]

b $C_3H_6(g) + 4\frac{1}{2} O_2(g) \rightarrow 3CO_2(g) + 3H_2O(g)$
[products = 1 mark]
[balancing = 1 mark]

4 a Unsaturated – has double bonds.
Hydrocarbons – only contain carbon and hydrogen atoms. [1]

b C_nH_{2n} [1] **c** C_8H_{16} [1]

5 a

C	H
$\frac{85.7}{12.0} = 7.14$	$\frac{14.3}{1.0} = 14.3$
$\frac{7.14}{7.14} = 1$	$\frac{14.3}{7.14} = 2$

b C_8H_{16}

6 B [1] **7** D [1]

Chapter 15

1 a

C	H	O
$\frac{68.2}{12.0} = 5.68$	$\frac{13.6}{1.0} = 13.6$	$\frac{18.2}{16.0} = 1.14$
$\frac{5.68}{1.14} = 5$	$\frac{13.6}{1.14} = 12$	$\frac{1.14}{1.14} = 1$

[2]

b $C_5H_{12}O$ [2]

2 a When H_2O is removed from a substance. [1]

b concentrated sulfuric acid / concentrated phosphoric acid / aluminium oxide [1]

c pent-1-ene [1]

3 a C_4H_9OH or $C_4H_{10}O$ [1]

b Structural isomers have the same molecular formula but a different structural formula. [1]

c Butan-1-ol is a primary alcohol and is oxidised. [1]
There is a colour change orange to blue / green. [1]
2-methylpropan-2-ol is a tertiary alcohol and is not oxidised. [1]
There is no colour change. [1]

4 a [structure: ethanal H-C(H)(H)-C(H)=O] [1]

b [structure: ethanoic acid H-C(H)(H)-C(=O)-OH] [1]

c [structure: propanone H-C(H)(H)-C(=O)-C(H)(H)-H] [1]

5 a hexanoic acid [1]
b i hexan-1-ol [1]
ii Primary. [1]

6 B [1]

Chapter 16

1 a carbon dioxide [1] **b** carboxylic acid [1]

2 a Contains carbon and hydrogen only. [1]

b Add bromine water and shake. [1]
The bromine water would decolourise if an alkene was present. [1]

3 a C=O [1]

b propanal [1]
propanone [1]
Both these compounds have the same molecular formula and both have a C=O bond which would give a peak at $1700\,\text{cm}^{-1}$. [1]

4 a A parent ion is formed when one electron is removed from the compound. [1]

b $C_{10}H_{16}O_4$
$(12 \times 10) + (1.007829 \times 16) + (15.99491 \times 4)$
$= 200.1049$ [2]

5 C [1] **6** A [1]

Answers to practice questions

Chapter 17
1. a A : $K^+(g) + \frac{1}{2}F_2(g)$ B : $K^+(g) + F(g) + e^-$
      C : LE (KF) [3]
   b $-929\,kJ\,mol^{-1}$ [2]
2. a $240\,000\,J$ [1]
   b $5.55\,mol$ [1]
   c $\Delta S = -(\Delta G - \Delta H)/T$ [1]
   d $\Delta S = \Delta H/T$ [1]
   e $373\,K$ [1]
   f $+117\,J\,K^{-1}\,mol^{-1}$ [1]
   g The sign is positive because steam is a more disordered state than liquid water. [1]
3. Lattice enthalpy is proportional to the square of the charge on each ion, and inversely proportional to the radius. Each ion in MgO has double the charge of those in NaF while the radii of the ions are roughly the same, because the elements are in the same period of the Periodic Table. [4]
4. a $\Delta S = -174.8\,J\,K^{-1}\,mol^{-1}$ [1]
   b Negative ΔS as expected because a gas changes into a solid. [1]

Chapter 18
1. a i 2   ii 1   iii 3 [3]
      Between experiments 1 and 2, [NO] is constant and [H_2] doubles. The rate doubles, so Rate $\propto$ [H_2]^1.
      Between Experiments 4 and 5, [H_2] is constant and [NO] doubles. The rate quadruples, so Rate $\propto$ [NO]^2. [1]
   b Rate = k [NO]^2 [H_2] [2]
   c $8.33 \times 10^4\,dm^6\,mol^{-2}\,s^{-1}$ [2]
   d $8.33 \times 10^{-5}\,mol\,dm^{-3}\,s^{-1}$ [2]
   e At this time, the concentrations of both reactants are known exactly. [1]
2. a 1; the units of the rate constant are s^{-1}. [2]
   b $89\,kJ\,mol^{-1}$. [10]
   c The value is realistic. It is of the same order of magnitude as the strength of covalent bonds. [2]
3. D [1]

Chapter 19 and 20
1. a $K_p = p^2SO_3(g)_{eqm} / p^2SO_2(g)_{eqm}\,pO_2(g)_{eqm}$ Units Pa^{-1} (or kPa^{-1}) [2]
   b Increase the total pressure or reduce the temperature. [2]
   c i Increase, ii no change. [2]
   NB units could equally well be expressed in kPa.
2. a i Pa; ii No units; iii Pa^{-2}. [3]
   b Pa^2 [1]
3. a $10 \times 10^4\,Pa$ [1]
   b 48.4 (no units) [2]
4. a Pb(s) | Pb^{2+} (aq) ∥ Cu^{2+}(aq) | Cu(s) $E^\ominus$ = +0.47 V [4]
   b Concentration of both solutions $1\,mol\,dm^{-3}$; temperature 298 K. [2]
   c Pb/Pb^{2+} to Cu/Cu^{2+} [1]
   d Pb(s) → Pb^{2+}(aq) + 2e^-
      Cu^{2+}(aq) + 2e^- → Cu(s) [2]
   e Pb(s) + Cu^{2+}(aq) → Pb^{2+}(aq) + Cu(s) [2]
   f The solution would gradually change from blue to colourless and a deposit of brown copper would form on the lead. [2]
5. Bromine will oxidise Fe^{2+} to Fe^{3+}. [1]
   $2Fe^{2+}$(aq) + Br_2(aq) → $2Fe^{3+}$(aq) + $2Br^-$(aq) [2]
   emf = 0.30 V (Fe^{3+}/Fe^{2+} more negative). [1]
6. [H^+(aq)] = $1\,mol\,dm^{-3}$; temperature 298 K; pressure of H_2 = 100 kPa; electrode of platinum [4]

Chapter 21
1. a 0.30 [1]
   b 0.70 [1]
   c 2.60 [2]
   d 11.00 [2]
   e 4.56 [2]
2. a $CH_3CH_2COOH(aq) \rightleftharpoons CH_3CH_2COO^-(aq) + H^+(aq)$ [1]
   b $K_a = \dfrac{[CH_3CH_2COO^-(aq)]_{eqm}\,[H^+(aq)]_{eqm}}{[CH_3CH_2COOH(aq)]_{eqm}}$ [1]
   c i 2.93 [2]
      ii 4.86 [2]
3. a $1.31 \times 10^{-3}\,mol\,dm^{-3}$
   b $K_a = 1.7 \times 10^{-5}\,mol\,dm^{-3}$ pK_a = 4.76 [6]

Chapter 22
1. a i $Na_2O(s) + H_2O(l) \rightarrow 2NaOH(aq) \rightarrow 2Na^+(aq) + 2OH^-(aq)$ [2]
      ii $SO_2(g) + H_2O(l) \rightleftharpoons H_2SO_3(aq) \rightleftharpoons H^+(aq) + HSO_3^-(aq)$ [2]
   b The trend is for basic oxides on the left of the period to acidic oxides on the right. [1]
   c Amphoteric means showing both acidic and basic behaviour. [1]
      Aluminium oxide will react with both acids and bases:
      $Al_2O_3(s) + 6HCl(aq) \rightarrow 2AlCl_3(aq) + 3H_2O(l)$ (basic behaviour) [2]
      $Al_2O_3(s) + 2NaOH(aq) + 3H_2O(l) \rightarrow 2NaAl(OH)_4$ (aq) (acidic behaviour) [2]

Answers to practice questions

2 a $P_4O_{10}(s) + 6H_2O(l) \rightarrow 4H_3PO_4(aq)$ [2]
  b i More than enough magnesium oxide to react with all the phosphoric acid. [1]
  +1 +5 −2   + 2 −2   +2 +5 −2   + 1 −2
  ii $2H_3PO_4(aq) + 3MgO(s) \rightarrow Mg_3(PO_4)_2(aq) + 3H_2O(l)$ [2]
  iii Oxidation states above. [2]
  No. There is no change in any oxidation state. [2]
  c It is insoluble and will sink to the bottom of its container. It can be filtered off. [2]
  d Sodium hydroxide is soluble and excess cannot be removed by filtration. So the waste water would be alkaline rather than neutral. [1]

3 a Al_2O_3 [1]
  b P_4O_{10} and SO_3 [2]
  c SiO_2 [1]
  d SO_3 [1]
  e i $Na_2O(s) + H_2O(l) \rightarrow 2NaOH(aq)$ [2]
  ii $SO_3(g) + H_2O(l) \rightarrow H_2SO_4(aq)$ [2]
  f $H_2SO_4(aq) + 2NaOH(aq) \rightarrow Na_2SO_4(aq) + 2H_2O(l)$ [2]

Chapter 23

1 a i $1s^2 2s^2 2p^6 3s^2 3p^6 3d^3$ [1]
  ii $1s^2 2s^2 2p^6 3s^2 3p^6 3d^{10}$ [1]
  b Chromium forms ions with part-filled d orbitals. Zinc ions have full d orbitals. [1]
  c i [1]

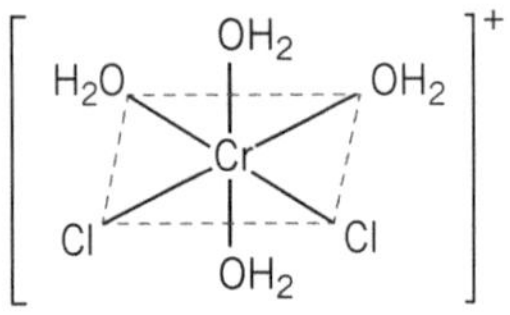

  [diagram of Cr complex with OH_2, H_2O, Cl ligands, charge +]

  ii E-Z (cis-trans) isomerism. [1]
  iii The E (trans) isomer. [1]

2 a $[Cu(H_2O)_6]^{2+}$ is octahedral; $CuCl_4^{2-}$ is tetrahedral [2].
  Cl^- is larger than H_2O and six Cl^- ions cannot fit around a Cu^{2+} ion [2].
  $[Cu(H_2O)_6]^{2+}$ has a Cu^{2+} ion and six neutral ligands. $CuCl_4^{2-}$ has a Cu^{2+} ion and four negatively charged ligands. [2]
  b Tetrahedral. Br^- is even bigger than Cl^-. [2]

3 a $[Ar]3d^9$ [1]
  b There is a space in the d orbitals, so electrons can move from one orbital to another. The ligands mean that there is an energy difference between the orbitals. When an electron moves from one orbital to one with a higher energy it absorbs electromagnetic energy in the visible part of the spectrum. [4]

  c i ΔE is the energy gap between orbitals, h is Planck's constant, v is the frequency of electromagnetic radiation. **ii** The frequency of electromagnetic radiation in the visible region is related to its colour. [4]

4 a The reaction starts slowly, speeds up, and then slows down again. Normally reactions start quickly and then slow down as the concentration of the reactants decreases. [2]
  b Autocatalytic – a product is a catalyst for the reaction. The reaction starts as normal but as the catalyst is produced it speeds up. It then slows down as the reactants are used up. [3]

Chapter 24

1 a The green precipitate is iron(II) carbonate. The brown precipitate is iron(III) hydroxide and the bubbles of gas are carbon dioxide. [2]
  b $Fe^{3+}(aq)$ is significantly more acidic than $Fe^{2+}(aq)$ and this makes the solution acidic enough to react with carbonate ions to form carbon dioxide. The Fe^{2+} solution is not acidic and the Fe^{2+} ions react with carbonate ions to form solid iron(II) carbonate. [2]
  c $[Fe(H_2O)_6]^{3+}(aq) \rightarrow [Fe(H_2O)_5OH]^{2+}(aq) + H^+(aq)$
  $2H^+(aq) + CO_3^{2-}(aq) \rightarrow CO_2(g) + H_2O(l)$
  $[Fe(H_2O)_6]^{2+}(aq) \; CO_3^{2-}(aq) \rightarrow FeCO_3(s) + 6H_2O(l)$ [6]

2 a A chelate is a ligand with the ability to form two or more co-ordinate bonds with a metal ion. [1]
  b $[Cu(H_2O)_6]^{2+} + EDTA^{4-} \rightarrow [Cu\,EDTA]^{2-} + 6H_2O$ [2]
  c i 2 **ii** 7 [2]
  d There is an entropy increase as the reaction proceeds from left to right because of the increased number of entities. [2]

3 a $[CuCl_4]^{2+}(aq)$ [1]
  b i $CuCO_3(s)$ **ii** Sodium carbonate solution (or other solution containing CO_3^{2-} ions). [2]
  c i $Cu(OH)_2$ (s) **ii** A deep blue solution. [2]
  d i Aqueous ammonia. **ii** A ligand displacement reaction. [2]
  e +2 [1]

Chapter 25

1 a Chiral means 'handed'. The molecule exists as two mirror image molecules that are not superimposable. [2]
  b One isomer may be an active drug while the other may be inactive or even toxic. Many syntheses produce a mixture of the two isomers which will be difficult to separate. [2]

Answers to practice questions

c [1]

d It is a carbon atom bonded to four different groups. [1]

2 a propan-1-ol, $CH_3CH_2CH_2OH$ [2]
  b propan-2-ol, $CH_3CH(OH)CH_3$ [2]
  c propanone, CH_3COCH_3 [2]

3 a 3 [1]
  b Prop- [1]
  c Carbon number 2. [1]
  d 2-aminopropanoic acid [1]
  e Optical isomerism. [1]
  f They rotate the plane of polarisation of polarised light in opposite directions. [1]

Chapter 26

1 a i pentanal [1]
    ii pentan-3-one [1]
    iii propanal [1]
  b i and ii [1]
  c i and iii [2]
  d ii [1]
  e ii [1]
  f sodium tetrahydridoborate(III) (sodium borohydride), $NaBH_4$ [1]
  g i and iii [2]

2 a i
  H–C(H)(H)–C(H)(H)–C(=O)–C(H)(H)–H [1]
   ii The C=O cannot be at the end of the chain and both the carbon atoms in the body of the chain are equivalent. [1]

3 a i Ethanol and methanoic acid, ii methanol and ethanoic acid. [4]
  b The sodium salts of the acids would be produced. [1]
  c i $HCOOC_2H_5 + H_2O \rightleftharpoons C_2H_5OH + HCOOH$ [2]
    ii $CH_3COOCH_3 + NaOH \rightarrow CH_3OH + CH_3COONa$ [2]

4 a $CH_3CHO + 2[H] \rightarrow CH_3CH_2OH$ [1]
  b Sodium tetrahydridoborate(III) generates the H^- ion, which has a lone pair of electrons and is a nucleophile. It will attack the $C^{\delta+}$ of the $C^{\delta+}=O^{\delta-}$ but it will not attack C=C, whose double bond is an area of high electron density. [2]

Chapter 27

1 a i bromobenzene [1]
    ii ethylbenzene [1]
    iii 1,4-dimethylbenzene [1]
  b ii and iii [1]
  c  [1]

2 The aromatic ring is attacked by electrophiles because of the high electron density of the delocalised electron system.

The reactions are substitutions because addition or elimination reactions would destroy the delocalised system and lose the extra stability of the delocalised ring system. [2]

3 a NO_2^+ [1]
  b $H_2SO_4 + HNO_3 \rightarrow NO_2^+ + HSO_4^- + H_2O$ [2]
  c benzene + NO_2^+ → nitrobenzene + H^+ [2]

  d i $C_6H_5NO_2 + 6[H] \rightarrow C_6H_5NH_2 + 2H_2O$ [2]
    ii Tin and hydrochloric acid. [1]

Chapter 28 and 29

1 a A primary amine has one organic group and two hydrogen atoms bonded to a nitrogen atom.

  A secondary amine has two organic groups and one hydrogen atom bonded to a nitrogen atom.

  A tertiary amine has three organic groups and no hydrogen atoms bonded to a nitrogen atom. [3]

  b The nitrogen has a lone pair of electrons that can accept a proton. [1]

  c A base uses its lone pair to form a bond with a proton (H^+ ion) and a nucleophile uses its lone pair to form a bond with an electron deficient carbon atom ($C^{\delta+}$). [1]

2 a propanenitrile, $CH_3CH_2C\equiv N$; bromopropane, $CH_3CH_2CH_2Br$ (or other halogenopropane) [4]

  b Propylamine can be synthesised from propanenitrile and hydrogen with a catalyst such as nickel. This is a reduction reaction. [3]

  Bromopropane may be reacted with excess ammonia to form propylamine. This is a nucleophilic substitution reaction. [3]

  c The reduction of propanenitrile. This gives only one product whereas the synthesis from a halogenoalkane risks the formation of secondary and tertiary amines as well as the desired product. [2]

Answers to practice questions

3 a The NH$_2$ group can form hydrogen bonds with water but the benzene ring cannot. [1]

   b On adding acid, the NH$_2$ group accepts proton to form an ionic compound C$_6$H$_5$NH$_3^+$Cl$^-$, which is water soluble. Adding alkali removes the proton to restore the sparingly soluble phenylamine. [2]

4 Addition polymers: the monomers have a C=C bond and link together so that the empirical formula of the polymer is the same as that of the monomer. [2]

   Condensation polymers have monomers with two functional groups which react together to eliminate a small molecule such as H$_2$O or HCl. [2]

5 a [1]

   —C—C—
     | |
     Cl H

   b Chloroethene,

   H   H
    \\ /
     C=C
    / \\
   H   Cl
   [2]

   c Addition polymerisation [1]

   d For example,

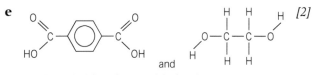

   (but start at any point, and stop when the same pattern of atoms reappears). [1]

   e [2]

   O   O   H H
   ‖   ‖ | |
   C—⌬—C   and O—C—C—O
   | | | |
   HO OH   H H

   (A diacid chloride would also be correct.)

   f Condensation polymer. [1]

   g The second one. It may be hydrolysed. [2]

6 Nylon has only two monomers, a diamine and a dicarboxylic acid (or diacid chloride).

   Proteins are poly-amino acids. Their monomers (amino acids) each have an amine group and a carboxylic acid group. Proteins are polymerised from up to 20 different amino acids, CH(NH$_2$)(R) COOH, which have different R groups. [4]

Chapter 30

1 a A formula drawn to show every atom and every bond. [2]

   b [2]

   c Alanine (ala) and glycine (gly). [4]
   d A zwitterion. [1]
   e i Boil with 6 mol dm^{-3} hydrochloric acid. [3]
      ii Water [1]
   f Thin-layer chromatography (TLC) or paper chromatography. [1]

   g i The COO$^-$ group would accept a proton.
      ii The NH$_3^+$ group would lose a proton. [2]

2 Hydrogen bonds; ionic bonds; sulfur–sulfur (disulfide) bonds. [3]

3 a 3 [1]
   b 2 [1]

4 a i [1]

   ii [1]

   b It exists as a zwitterion in which the amine group is protonated and the carboxylic acid deprotonated, so it is essentially ionic. [1]

   c It has no carbon atom with four different groups bonded to it. All the others do. [1]

   d Aminoethanoic acid. [1]

Chapter 31

1 a ethanol [1]
   b bromoethane (or other halogenoethane) [1]
   c propan-2-ol [1]
   d nitrobenzene [1]

2 For **a** react ethene with concentrated sulfuric acid followed by addition of water to give ethanol. Then oxidise the ethanol by refluxing with excess potassium dichromate solution acidified with concentrated sulfuric acid. [2]

   For **d** react benzene with a mixture of concentrated nitric and sulfuric acids to give nitrobenzene. This can be reduced to phenylamine using tin and hydrochloric acid. [2]

3 Step 1 Dehydrate propan-1-ol to propene. Step 2 Use concentrated sulfuric acid followed by water to produce propan-2-ol. Step 3 Oxidise the propan-2-ol to propanone with acidified potassium dichromate. [3]

4 a 1-chlorobutane (step 1) → butan-1-ol (step 2) → butanoic acid

      Step 1: Warm [1] with aqueous sodium (or potassium) hydroxide [1] with ethanol as a co-solvent. [1]

      Step 2: Reflux [1] with excess [1] potassium dichromate solution [1] with concentrated sulfuric acid catalyst. [1]

5 a React with CN$^-$ ions [1]
   b Butanenitrile, CH$_3$CH$_2$CH$_2$CN [2]
   c i Hydrogen with a nickel catalyst [2]
      ii Reaction with ammonia will produce some secondary and tertiary amines and quaternary ammonium salts as well as the target molecule. [2]

Answers to practice questions

6 a alkene and carboxylic acid [2]
   b Alkene – shake with bromine solution, the red-brown colour disappears [2]
      Carboxylic acid – add sodium (or potassium) carbonate solution, the mixture fizzes [2]
   c $CH_2BrCHBrCH_2COOH$, $CH_2=CHCH_2COONa$ [2]

Chapter 32

1 a 2 [1]
   b The hydrogen atoms are not on adjacent carbon atoms. [1]
   c Tetramethylsilane (TMS). This is added to the sample to calibrate the spectrum. [2]
   d CCl_4 or CCl_3D or D_2O. The solvent must not contain hydrogen atoms which would swamp the signal from the hydrogens in the sample. [2]

2 a Three, because there are three different types of environment for the carbon atoms; C=O, CH_2, and CH_3, which differ in how far away they are from the electronegative oxygen atom. [2]
   b Two, because there are just two different environments of hydrogen atoms, CH_2 and CH_3. [2]
   c The CH_3 peaks and the CH_2 peaks will be in the ratio 3 : 2. [1]
   d The peak corresponding to CH_2 will be split into 4 (height ratio 1 : 3 : 3 : 1) by the three adjacent hydrogens on the CH_3 group$_2$. That due to CH_3 will be split into three (height ratio 1 : 2 : 1) by the two adjacent hydrogens on the CH_2 groups. [4]
   e The n+1 rule. [1]

3 a All the carbon atoms are in exactly the same enviroment. [1]
   b Tetramethylsilane (TMS), $(CH_3)_4Si$ [2]
   c Parts per million (ppm). [1]
   d None, there are no hydrogen atoms. [2]

Chapter 33

1 a i R_f = distance moved by spot / distance moved by solvent.
     ii A 0.56; B 0.36; C 0.90 [4]
   b i B;
     ii the solvent in which the chromatogram is run. [2]
   c Shine ultraviolet light on the plate. Spray the plate with a chemical that reacts with amino acids to give a coloured product. [2]
   d The spot cannot move further than the solvent. [1]

2 [4]

	Stationary phase	Mobile phase
Thin-layer	Silica or alumina	Liquid solvent
Column	Silica or alumina or resin	Liquid solvent
GLC	Oil coated on powder (or on the inside of a capillary tube)	Unreactive gas

3 Place a suitable solvent (the eluant) in a beaker to a depth of about 0.5 cm.

Mark a line about 1 cm from the bottom of a TLC plate and place a small spot of the amino acid mixture on this line.

Place the TLC plate in the solvent and place a lid on the beaker.

Allow the solvent to climb up the plate (by capillary action) until the solvent has almost reached the top of the plate.

Mark the position reached by the solvent (the solvent front).

Remove the plate from the beaker and locate the position of the spots using UV light or a suitable chemical spray. *[6 marks]*

193

Answers to summary questions

1.1/1.2

1. a 11 protons, 12 neutrons, 11 electrons [1]
   b 8 protons, 8 neutrons, 10 electrons [1]
   c 35 electrons, 45 neutrons, 35 electrons [1]
   d 19 protons, 21 neutrons, 18 electrons [1]
2. They have the same number of electrons/same electron arrangement. [1]
3. Protons have a relative mass of 1 and relative charges of +1, neutrons have a relative mass of 1 and no charge. [2]

1.3

1. A circle/sphere. [1]
2. a $1s^2 2s^2 2p^6 3s^1$ [1]
   b $1s^2 2s^2 2p^6 3s^2 3p^4$ [1]
   c $1s^2 2s^2 2p^6 3s^2 3p^5$ [1]
   d $1s^2 2s^2$ [1]
3. a $1s^2 2s^2 2p^6$ [1]
   b $1s^2 2s^2 2p^6 3s^2 3p^6$ [1]
   c $1s^2 2s^2 2p^6$ [1]
   d $1s^2 2s^2 2p^6$ [1]

1.4

1. Vacuum, ionisation, acceleration, ion drift, detection, data analysis. [1]
2. 20.179 to three significant figures the answer is 20.2. $(20.0 \times 0.909) + (21.0 \times 0.003) + (22 \times 0.088) = 20.179$ to three significant figures the answer is 20.2. [2]

1.5

1. $Mg^{2+}(g) \rightarrow Mg^{3+}(g) + e^-$ [2]
2. Although there is an increase in nuclear charge down the group, the electrons in the outer shell are further from the nucleus and there is more shielding so the outer electron is lost more easily down the group. [2]
3.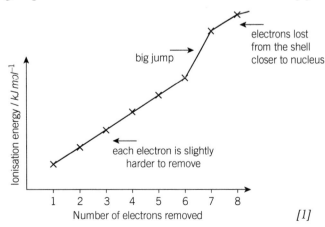

2.1

1. a $\dfrac{5.61}{56.1} = 0.100$ moles [1]
   b $0.150 \times 39.1 = 5.87\,g$ [1]
   c $0.32 \times 23.9 = 7.6\,g$ [1]

2.2

1. $2.00\,mol\,dm^{-3}$ [1]
2. $n(HNO_3) = 0.00414$ [1]
   $n(NaOH) = 0.00414$ [1]
   $0.166\,mol\,dm^{-3}$ [1]

2.3

1. Vol $H_2(g) = 90.0\,cm^3$
   Ratio $H_2 : NH_3 = 3 : 2$
   Vol of $NH_3(g)$ formed $= \dfrac{90.0}{3 \times 2} = 60.0\,cm^3$ [1]
2. $V = \dfrac{nRT}{p} = \dfrac{2.31 \times 8.31 \times (32 + 273)}{2 \times 10^5} = 0.029\,m^3$ [2]
3. $n(Mg) = \dfrac{mass}{M_r} = \dfrac{0.2}{24.3} = 8.23 \times 10^{-3}$
   ratio $Mg : H_2 = 1 : 1$
   $n(H_2) = 8.23 \times 10^{-3}$
   $V = \dfrac{nRT}{p} = \dfrac{(8.23 \times 10^{-3}) \times 8.31 \times 298)}{1 \times 10^5}$ [3]
   $= 2.04 \times 10^{-4}\,m^3$

2.4/2.5

1. $2Na(s) + Cl_2(g) \rightarrow 2NaCl(s)$ [2]
2. Fe $\dfrac{69.9}{55.8} = 1.25$ O $\dfrac{30.1}{16} = 1.88$
   $\dfrac{1.25}{1.25} = 1$ $\dfrac{1.88}{1.25} = 1.505$ double everything [3]
   2   3
   Fe_2O_3
3. $\dfrac{42.0}{14} = 3$ so the molecular formula = C_3H_6 [2]

2.6

1. % yield $= \left(\dfrac{actual\ yield}{theoretical\ yield}\right) \times 100$
   n ethene reacting $= \dfrac{2.00}{28.0} = 0.0700$ moles theoretical
   moles of bromoethane $= 0.0700$
   theoretical yield of bromoethane $= 0.0700 \times 108.9$
   $= 7.63\,g$ [3]
   % yield $= \left(\dfrac{5.80}{7.63}\right) \times 100 = 76.0\%$
2. atom economy $= \dfrac{relative\ mass\ of\ desired\ product}{total\ relative\ mass\ of\ products} \times 100$
   $= \dfrac{35.5 \times 2}{2(23.0 + 35.5) + 71.0 + 2} \times 100$
   $= \dfrac{71.0}{190.0} \times 100 = 37.4\%$ [2]

Answers to summary questions

3 The percentage yield is used to compare the actual yield of a reaction with the theoretical yield of the reaction. The atom economy is the proportion of reactants that are converted into useful products. In this case the atom economy is 100% as it is an addition reaction. [2]

3.1

1   a NaBr [1]
    b $Ca(OH)_2$ [1]
    c $MgCl_2$ [1]
2 When molten the ions can move but when solid the ions cannot move. [1]

3.2

1 A shared pair of electrons. [2]
2 A shared pair of electrons but where both electrons come from the same atom. [2]
3 Nitrogen is a simple molecule so there are only weak forces of attraction between the molecules. [2]

3.3

1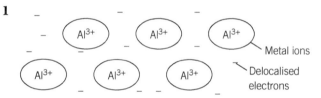

The delocalised electrons can move. [2]
2 Sodium, magnesium, aluminium. Across the period the strength of the metallic bonding increases. [2]
3 In metals the layers of cations can slip over each other if a large enough force is applied. The strength of the metallic bond stops the attractions being broken completely. [1]

3.4

1   a $\overset{\delta+}{H}-\overset{\delta-}{F}$ [1]
    b $\overset{\delta+}{C}-\overset{\delta-}{Cl}$ [1]
    c $\overset{\delta-}{O}-\overset{\delta+}{N}$ [1]
    d $\overset{\delta+}{S}=\overset{\delta-}{O}$ [1]
2   a $\overset{\delta+}{H}-\overset{\delta-}{Br}$ Polar [2]
    b $\overset{\delta+}{H}-\overset{\delta-}{S}$ Polar [2]
    c 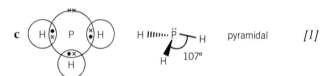 Polar [2]
    d 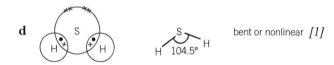 Polar [2]
3   a $NaCl, MgCl_2, AlCl_3$ [2]
    b NaCl, NaBr, NaI [2]

3.5

1 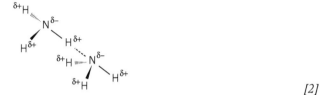 [2]

2 Ethane. It has more electrons therefore stronger van der Waals' forces so more energy is needed to overcome the forces of attraction. [2]

3.6

1   a trigonal planar [1]
    b tetrahedral [1]
    c pyramidal [1]
    d bent or nonlinear [1]

3.7

1

Name of substance	Formula	Type of structure	Type of bonding
magnesium	Mg	giant metallic	metallic
sodium chloride	NaCl	giant ionic	ionic
chlorine	Cl_2	simple molecular	covalent
graphite	C	giant covalent/ macromolecular	covalent

[1]

2 Cl_2 – there are only weak van der Waals' forces between molecules.

H_2O – there are hydrogen bonds between molecules.

$MgCl_2$ – there are strong ionic bonds between the ions. [2]

4.1/4.2

1 Zero. Nitrogen is a diatomic gas at room temperature. [1]
2 $C_2H_5OH(l) + 3O_2(g) \rightarrow 2CO_2(g) + 3H_2O(g)$ [2]

195

Answers to summary questions

4.3

1. Heat loss. Use a polystyrene cup or add insulation to the beaker. [2]
2. a Zinc powder has a high surface area so reacts quickly. [1]
   b $50.0 \times 4.18 \times (30.5 - 21.0) = 1985.5$ J or 1.9855 kJ mol^{-1} [2]
   c moles $= \dfrac{50.0}{1000} \times 1.00 = 0.05$ mol
   enthalpy change $= \dfrac{1.9855 \text{ kJ}}{0.05 \text{ mol}} = -39.7$ kJ mol^{-1}
   The negative sign shows the reaction is exothermic. [2]

4.4

1. The enthalpy change of a chemical reaction is independent of the route by which the reaction is achieved and depends only on the initial and final states. [1]
2. 100 kPa and a stated temperature of 25 °C or 298 K. [2]
3. $3C(s) + 3H_2(g) \rightarrow C_3H_6(g)$ [2]
4. $-(-791) + (-602) + (2 \times -33)$
   $= +123$ kJ mol^{-1} [2]

4.5

1. The enthalpy change when one mole of a compound is completely burnt in oxygen under standard conditions. [1]
2. $C_3H_6(g) + 4\tfrac{1}{2} O_2(g) \rightarrow 3CO_2(g) + 3H_2O(l)$ [1]
3. $2C(s) + 2H_2(g) \rightarrow C_2H_4(g)$
   ↓   ↓   ↓
   Combustion products
   $+(2x -394) + (2x -286) - (-890)$
   $= -470$ kJ mol^{-1} [3]

4.6

1. Zero / 0 kJ mol^{-1} [1]
2. Different forms of the same element that occur in the same state/phase. [1]
3. a

   $2C_2H_6(g) + 7O_2(g) \xrightarrow{\Delta H^\ominus} 4CO_2(g) + 6H_2O(g)$

   -172 kJ mol^{-1} ↖   ↗ -3292 kJ mol^{-1}

   $4C(s) + 6H_2(g) + 7O_2(g)$

   $\Delta H^\ominus = -(-172) + (-3292) = -3120$ kJ mol^{-1} [1]

   b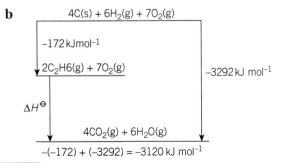

   $-(-172) + (-3292) = -3120$ kJ mol^{-1} [2]

4.7

1. The mean (average) bond enthalpy is the mean amount of energy required to break one mole of a specified type of covalent bond in a gaseous species. [1]
2. $H-Cl(g) \rightarrow H(g) + Cl(g)$ [1]
3. The mean bond enthalpy for a bond is an average value obtained from many molecules. [1]
4. $\dfrac{1664 \text{ kJ mol}^{-1}}{4} = +416$ kJ mol^{-1} [1]

5.1/5.2

1. It is the minimum collision energy that particles must have to react. [1]
2. It increases the energy of the molecules. [1]
3. The average energy of the particles decreases so the peak of the distribution curve moves to the left as the average energy decreases. The distribution curve becomes higher because the total number of particles remains the same. [2]
4. Particles collide more often and more particles have enough energy to react. [1]

5.3

1. A catalyst increases the rate of reaction but does not affect the yield of the reaction. [2]
2.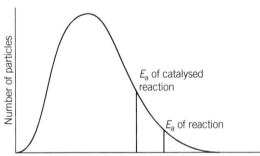

   Adding a catalyst decreases the activation energy of the reaction, a greater proportion of the particles now have enough energy to react so the rate of reaction increases. [2]

6.1/6.2

1. The reaction is reversible. [1]
2. When the conditions of a dynamic equilibrium are changed then the position of equilibrium will shift to minimise the change. [1]
3. It increases the rate of reaction by offering an alternative reaction pathway with a lower activation energy.
   It does not affect the position of equilibrium as the rate of forwards reaction and the rate of backwards reaction are both increased by the same amount. [2]

6.3

1. Where nothing can enter or leave. [1]
2. Iron, it increases the rate of reaction but does not affect the yield of ammonia. [3]

6.4

1. **a** $K_c = \dfrac{[NO(g)]^2\,[O_2(g)]}{[NO_2(g)]^2}$ [1]

   b $K_c = \dfrac{[H_2(g)]\,[Cl_2(g)]}{[HCl(g)]^2}$ [1]

   c $K_c = \dfrac{[SO_3(g)]^2}{[SO_2(g)]^2\,[O_2(g)]}$ [1]

2. **a** $\dfrac{(\cancel{\text{mol dm}^{-3}} \times \cancel{\text{mol dm}^{-3}}) \times \text{mol dm}^{-3}}{(\cancel{\text{mol dm}^{-3}} \times \cancel{\text{mol dm}^{-3}})} = \text{mol dm}^{-3}$ [1]

   b $\dfrac{(\cancel{\text{mol dm}^{-3}} \times \cancel{\text{mol dm}^{-3}})}{(\cancel{\text{mol dm}^{-3}} \times \cancel{\text{mol dm}^{-3}})} = \text{no units}$ [1]

   c $\dfrac{(\cancel{\text{mol dm}^{-3}} \times \cancel{\text{mol dm}^{-3}})}{(\cancel{\text{mol dm}^{-3}} \times \cancel{\text{mol dm}^{-3}}) \times \text{mol dm}^{-3}} = \text{mol}^{-1}\,\text{dm}^3$

   $= \text{dm}^3\,\text{mol}^{-1}$ [1]

6.5

1. $\dfrac{K_c\,[HCl(g)]^2}{[H_2(g)]\,[Cl_2(g)]} = \dfrac{(0.800)^2}{0.0500 \times 0.100} = 128$ [3]

   No units

2.

	$N_2O_4(g)$	$NO_2(g)$
initial amount (mol)	0.600	0.000
amount at equilibrium (mol)	0.250	0.700
concentration at equilibrium (mol dm^{-3})	0.500	1.40

$K_c = \dfrac{[NO_2(g)]^2}{[N_2O_4(g)]}$

$= \dfrac{(1.40)^2}{0.500}$

$= 3.92\,\text{mol dm}^{-3}$ [5]

6.6

1. A decrease in pressure will shift the position of equilibrium to the left-hand side as it is the side with the more gaseous molecules. [2]

2. An increase in temperature will shift the position of equilibrium to the left-hand side as this is the endothermic direction. [2]

3. Increasing the temperature will decrease the value of the equilibrium constant, K_c as the reaction is exothermic in the forward direction. [2]

7.1/7.2

1. 0 [1]
2. +6 [1]
3. **a** +5 [1]
   b −1 [1]
   c +7 [1]

7.3

1. Oxidation is the loss of electrons. Reduction is the gain of electrons. [2]
2. $Fe^{2+} + Zn \rightarrow Fe + Zn^{2+}$ [1]
3. $Zn \rightarrow Zn^{2+} + 2e^-$ [1]
   $2H^+ + 2e^- \rightarrow H_2$ [1]
4. **a** $Mg + Cl_2 \rightarrow MgCl_2$ [1]
   b $Mg \rightarrow Mg^{2+} + 2e^-$ [2]
   $Cl_2 + 2e^- \rightarrow 2Cl^-$ [2]

8.1/8.2

a $1s^2 2s^2 2p^1$ p-block [1]
b $1s^2 2s^2 2p^5$ p-block [1]
c $1s^2 2s^2 2p^6$ p-block [1]
d $1s^2 2s^2 2p^6 3s^2 3p^6 3s^1$ s-block [1]
e $1s^2 2s^2 2p^6 3s^2 3p^6 4s^2 3d^6$ d-block [1]

2. Regularly recurring. [1]
3. **a** Covalent, simple molecules. [1]
   b Metallic, giant metallic. [1]
4. Sulfur forms simple molecules. When sulfur melts only the weak van der Waals' forces between molecules are broken. This requires little energy and happens at low temperatures. [1]

   Silicon forms a giant covalent structure. When silicon melts the strong covalent bonds between atoms are broken. A lot of energy is required and this only happens at high temperatures. [1]

8.3/8.4

1. First ionisation energy is the energy required to remove 1 electron from each atom in 1 mole of gaseous atoms forming 1 mole of ions with a single positive charge. [2]

2. Decrease. Each atom has one more proton in its nucleus so is able to attract its outer electron more strongly. [2]

3. $Mg(g) \rightarrow Mg^+(g) + e^-$ [2]

4. The outer electron configuration of sulfur has a pair of electrons in one of the p orbitals. There is repulsion between the electrons so less energy is needed to remove an electron from this pair so sulfur has a lower first ionisation energy than phosphorus. [2]

9.1

1. The solubility of Group 2 sulfates decreases down the group. [1]

2. The atomic radii of the elements increase down the group.

   This is because down the group the atoms have another shell of electrons. [1]

3. The melting points generally decrease down the group.

Answers to summary questions

Down the group the size of the metal ions increases so the strength of the metallic bonding decreases.

Less energy is required to overcome the forces of attraction so they melt at lower temperatures. *[1]*

10.1

1. Electronegativity is a way of measuring the attraction that a bonded atom has for the electrons in a covalent bond. *[1]*
2. Increases down the group. Down the group atoms have an extra shell of electrons and become larger. *[2]*
3. Decreases down the group. Although atoms have more protons down the group, the atoms have more shells of electrons. This means there is more shielding and less attraction between the nucleus and the electrons. In addition, down the group the atomic radius increases. This results in less attraction between the nucleus and the electrons in the covalent bond. *[2]*

10.2

1. An oxidising agent is a species which oxidises another substance by removing electrons from it. *[1]*
2. The oxidising ability of the halogens decreases down the group.

   Down the group the electron which is gained is being placed into a shell which is further from the nucleus. Also the amount of shielding increases. As a result the attraction between the nucleus and the electron decreases so there oxidising ability goes down. *[3]*
3. i $Cl_2(g) + 2I^-(aq) \rightarrow 2Cl^-(aq) + I_2(aq)$ *[1]*

   ii Violet *[1]*

   iii The iodide, I^- ions (oxidation number −1) are oxidised to iodine, I_2 (oxidation number 0). *[2]*

10.3/4

1. a −1 *[1]*

   b +5 *[1]*
2. Add silver nitrate solution and a cream precipitate would form. *[2]*
3. Down the group the halide ions become increasingly good reducing agents. Down the group the atoms get larger so it becomes easier for an electron to be lost. *[2]*

11.1

1. a Displayed formula. *[1]*

   b C_4H_8 *[1]*

   c $CH_3CH_2CHCH_2$ *[1]*

   d CH_2 *[1]*

   e (structure) *[1]*

11.2

1. a hexane *[1]*

   b pentane *[1]*

   c but-2-ene *[1]*

   d 3-ethyl-hexane *[1]*

   e 2-chlorobutane *[1]*

   f 2-chloro-pentane *[1]*

11.3

1. Positional, functional, chain. *[1]*
2. butane / methylpropane *[2]*
3. 1-iodo-butane / 2-iodo-buane *[2]*

12.1

1. A compound that only contains carbon and hydrogen atoms and only has single covalent bonds. *[2]*
2. C_nH_{2n+2} *[1]*
3. a ethane *[1]*

   b propane *[1]*

   c pentane *[1]*
4. a hexane *[1]*

   b octane *[1]*
5. The boiling point increases as the number of carbon atoms increases. The larger the molecule is the stronger the van der Waals' forces between molecules. *[2]*
6. Propane is an alkane so is non-polar. This means it will not dissolve in polar solvents like water. *[1]*

12.2

1. When the fossilised remains of dead plants and animals are subjected to high temperatures and pressures. *[1]*
2. When burnt sulfur reacts with oxygen to make sulfur dioxide which causes acid rain. *[1]*
3. A part of the crude oil which contains alkane molecules with a similar number of carbon atoms and therefore a similar boiling point. *[1]*
4. The crude oil is heated until it vaporises. It then moves up the fractionating tower and cools. Each fraction condenses at a different point up the fractionating column where it is collected. Bitumen is collected at the bottom of the tower while refinery gases are collected at the top of the tower. *[3]*

12.3

1. a C_3H_6 *[1]*

   b C_7H_{16} *[1]*

   c $C_{18}H_{38}$ *[1]*

Answers to summary questions

2 a [diagram: H-C-C-C-C-C-C-H (hexane) → H-C-C-C-C-H (butane) + CH₂=CH₂ (ethene)] [1]

b [diagram: H-C-C-C-C-C-C-C-C-H (octane) → H-C-C-C-C-C-H (pentane) + CH₂=CH-CH₃ (propene)] [1]

12.4

1. A substance that can be burnt to release energy. [1]
2. When a substance is burnt in a large excess of oxygen. [1]
3. $C_4H_{10} + O_2 \rightarrow 4CO_2 + 5H_2O$ [2]
4. When the sulfur is burned it reacts with oxygen to form sulfur dioxide. This reacts with water to form first sulfurous acid and then sulfuric acid and this causes acid rain. [1]

12.5

1. A species with an unpaired electron. [1]
2. Free radical substitution. [2]
3. **a** UV light [1]
   b i $Cl_2 \rightarrow 2Cl$
   ii $Cl + C_2H_6 \rightarrow C_2H_5 + HCl$
   $C_2H_5 + Cl_2 \rightarrow C_2H_5Cl + Cl$ [2]
   iii $2Cl \rightarrow Cl_2$
   or $Cl + C_2H_5 \rightarrow C_2H_5Cl$
   or $2C_2H_5 \rightarrow C_4H_{10}$ [2]

13.1

1. **a** 2-bromobutane [1]
   b 2,2-dichloro-4-methylhexane [1]
2. The bond polarity of the C–X bond decreases down the group as the difference in electronegativity between the halogen and the carbon decreases down the group. [2]
3. The bond enthalpy of the C–X bond decreases down the group as smaller atoms attract the shared pair of electrons in the C–X bond more strongly. [2]

13.2

1. Nucleophiles have a lone pair of electrons which they can donate to another molecule to form a new covalent bond. [1]
2. Solvent. [1]

3. [mechanism diagram: CH₃-CHCl-CH₃ + :C≡N⁻ → CH₃-CH(CN)-CH₃ + :Cl⁻] [4]

13.3

1. In elimination reactions a small molecule is lost from an organic molecule. [1]
2. propan-2-ol [1]
3. [three structures: pent-1-ene, Z-pent-2-ene, E-pent-2-ene] [3]

14.1

1. A species that can accept a pair of electrons to form a new covalent bond. [1]
2. Electrophilic addition. [1]
3. [three structures: but-1-ene, Z-but-2-ene, E-but-2-ene] [3]

Answers to summary questions

14.2

1. Electrophilic substitution. [1]
2. 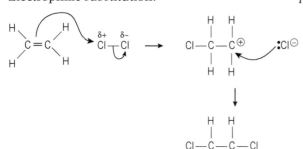 [4]
3. 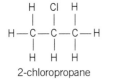 and 1-chloropropane [2]

14.3

1. a poly(ethene) [1]
   b polychloroethene [1]
   c polypropene [1]

2. $\left(\begin{array}{cc} H & CH_3 \\ -C-C- \\ H & Cl \end{array} \right)_n$ [1]

3. $\begin{array}{cc} H & Br \\ C=C \\ Cl & H \end{array}$ [1]

15.1

1. $C_nH_{2n+1}OH$. [1]
2. Secondary. [1]
3. There are hydrogen bonds between the water and methanol molecules.

   [2]

15.2

1. Fermentation and hydration of ethene. [1]
2. Fermentation. The sugar from plants, which are living things, that can be replanted. [1]
3. a 100% [1]
   b 51% [1]

15.3

1. Ketone [1]
2. Tollen's reagent (ammoniacal silver nitrate) or Fehling's reagent. [2]
3. Concentrated sulfuric acid or phosphoric acid. [1]
4. a propanoic acid. [1]
   b $CH_3CH_2CH_2OH + 2[O] \rightarrow CH_3CH_2COOH + H_2O$ [2]

16.1

1. It would decolourise bromine water [1]
2. Carbon dioxide is produced so it is a carboxylic acid. [1]
3. Tertiary [1]
4. Aldehyde [1]
5. Warm with aqueous sodium hydroxide. Add an acidified silver nitrate. A white precipitate of AgCl is seen. [2]

16.2

1. Faster, more accurate, more sensitive, and smaller samples can be used. [1]
2. It measures the relative atomic mass of the sample very precisely / to lots of decimal places. [1]
3. Relative molecular mass. [1]

16.3

1. Although both compounds are alcohols and have the same functional group (O–H) and would show absorptions in the same areas the fingerprint region can be used to tell them apart. [1]
2. Absorption peaks that should not be present would indicate that a sample contains impurities. [1]
3. The peak at 1700 cm^{-1} is caused by a C=O bond which occurs in aldehydes, ketones, carboxylic acids, and esters. [1]

   The broad peak between 2600 and 3500 cm^{-1} is caused by an O–H bond which occurs in carboxylic acids. [1]

   So the substance must be a carboxylic acid. As it has 3 carbon atoms it is propanoic acid. [1]

# Answers to summary questions

17.1/17.2

1. The first electron affinity involves the addition of an electron to the outer shell of the oxygen atom to which is being *attracted* by the positive charge of the nucleus. The second electron affinity involves the addition of an electron to a negatively charged ion which *repels* it. [2]

2. a $Mg^{2+} \rightarrow Mg^{3+} + e^-$ [1]
   b The first two electrons are removed from the outer shell of the magnesium atom. The third is removed from the second shell which is closer to the nucleus. [1]
   c This is the sum of the first three ionisation energies:
   738 + 1451 + 7733 = 9922 kJ mol⁻¹ [1]
   d This is the sum of the atomisation energy and the first ionisation energy of magnesium.
   147.1 + 738 = 885.1 kJ mol⁻¹ [2]

3. Step 0: Elements in their standard states. Step 1: Ca(s) → Ca(g). Step 2: S(s) → S(g). Step 3: Ca(g) → Ca⁺(g) + e⁻. Step 4: Ca⁺(g) → Ca²⁺(g) + e⁻. Step 5: S(g) + e⁻ → S⁻(g). Step 6: S⁻(g) + e⁻ → S²⁻(g) [7]

4. 3013 kJ mol⁻¹ [2]

5. The dip is explained by the fact that the first electron affinity of sulfur is negative and second electron affinity is positive. [2]

17.3

1. The oxygen atom, as it will have a δ– charge. [2]
2. +24 kJ mol⁻¹ [2]
3. The lithium ion is smaller than the sodium ion and the water molecules can get closer the lithium ions, giving out more energy. [1]
4. $AlBr_3$ because the aluminium ion has the greatest positive charge, which will polarise the bromide ion most. [2]

17.4

Go further

a −204 kJ mol⁻¹
b Bond energies are averages over a number of molecules.
c It will tend to be stable.
d The entropy change for the reaction.
e It will be negative because two gases react to form a solid.

Summary questions

1. a ΔS = −284 J K⁻¹ mol⁻¹ [1]
   b Negative because two gases (highly disordered) form a solid (highly ordered). [1]
2. Approximately zero because all the species are solids and each will have approximately the same (low) value for its entropy. [2]

3. a −69.5 kJ mol⁻¹ [1]
   b It is feasible at room temperature. [1]
   c Nothing. [1]
4. a ΔG = +14 kJ mol⁻¹ [1]
   b No, because ΔG is positive. [2]
5. a +197 kJ mol⁻¹, not feasible [2]
   b −208 kJ mol⁻¹, feasible [2]

18.1

1. a A product, as its concentration increases with time. [1]
   b The initial rate. [1]
   c Zero. [1]
2. a Use a colorimeter to measure the concentration of $CuSO_4$, which is blue. [1]
   b Sample the reaction mixture at intervals and measure the concentration of ethanoic acid by titrating it with an alkali. [1]
   c Collect the hydrogen gas in a syringe or inverted measuring cylinder of water. [1]

18.2

1. a i the rate constant [1]
      ii 3 [1]
      iii 2 [1]
      iv zero [1]
      v 1 [1]
   b A catalyst [1]
   c $dm^6 mol^{-2} s^{-1}$ [1]
2. a i It doubles ii it trebles. [2]
   b i It quadruples ii it increases ninefold. [2]
   c i No change ii no change. [2]

18.3

1. This is the only point in a reaction when the concentrations of all the species involved are known *exactly*. [1]
2. a First order with respect to X, second order with respect to Y. [2]
   b Rate = $k [X][Y]^2$ [1]
   c $mol^2 dm^6 s^{-1}$ [1]
   d No, there may be other species involved that have not been tested. [2]

18.4

1. a i The height decreases ii the peak moves to the right. [2]
   b The total number of molecules. [1]
   c The number of molecules with sufficient energy to react. [1]
   d Not at all. [1]

Answers to summary questions

2 a 103 kJ mol^{-1} [2]
 b It is the right order of magnitude comparable with bond energies. [1]

18.5

1 a A and B. [1] **b** C. [1] **c** Step 1. [1]
 d A catalyst.
 It is used up in step 1 and regenerated in step 2. [1]
 e W, Y, and B. [1]
 They appear after the rate-determining step. [1]

2 a 2 NO +N̶₂O̶₂ +H₂ +N̶₂O̶ +2H₂ → N̶₂O̶₂ +N̶₂O̶ + H₂O + N₂ +H₂O [1]
 b N_2O_2 and N_2O [2]
 c Yes, because both NO and H_2 appear in or before the rate-determining step. [2]

19.1

1 CO_2 20 kPa, O_2 40 kPa, N_2 40 kPa [3]
2 a $K_p = p^2 NO_2(g)_{eqm}/ pN_2O_4(g)_{eqm}$ [1] Units Pa [1]
 b No change. [1]
 K_p is constant unless the temperature changes. [1]
 c Move to left. [1]
 Le Chatelier's principle states that the equilibrium will move so as to restore the original pressure so it will move to the side with fewer molecules. [1]

20.1

1 a 0.63 V [1] **b** Pb positive [1]
 c Zn(s)|Zn^{2+}(aq) ∥ Pb^{2+}(aq)|Pb(s) $E^\ominus$ = +0.64 V [2]
2 a i A phase boundary (in this case between a solid electrode and a solution). [1]
 ii A salt bridge. [1]
 b The most oxidised species (Al^{3+} and Pb^{2+}) should be drawn next to the salt bridge [1], the sign of the emf is that of the right hand electrode as drawn. [1]

20.2

1 No, electrons flow from Fe^{3+}/Fe^{2+} to ½ Br$_2$/Br$^-$. [2]
2 a A Solution of Cu^{2+}(aq) ions of concentration 1 mol dm^{-3}
 B Copper rod
 C Voltmeter with the positive terminal connected to the copper rod
 D Salt bridge
 E Hydrogen gas at pressure 100 kPa
 F Platinum electrode coated with finely-divided platinum
 G Solution of H$^+$ ions of concentration 1 mol dm^{-3} [7]
 b 298 K [1]

20.3

1 A Water out [1]
 B Hydrogen in [1]
 C Oxygen in [1]
 D Electrolyte – sodium hydroxide [1]
 E Positive electrode (anode) [1]
 F Negative electrode (cathode) [1]

2

Cell	Advantage	Disadvantage
Daniell		Not portable due to liquid electrolyte
Leclanché	Solid electrolyte, therefore portable	May leak if zinc case is used up
Lead–acid	Rechargeable, high voltage	Contains an acidic liquid electrolyte
Fuel cell	Water is the only product, which may be useful in space	H_2 and O_2 are gases which are difficult to store, the hydrogen may come from fossil fuels

Other suggestions are possible. *[1 mark each]*

3 There are no harmful products. The water produced can be used for drinking. [2]

21.1

1 Acid, H_2SO_4, base, HNO_3 [2]
2 a NO_3^-, **b** Cl^-, **c** OH^-, **d** SO_4^{2-} [4]
3 10^{-12} mol dm^{-3} [2]

21.2

1 a 3; **b** 6; **c** 10; **d** 5.30 [4]
2 0.70 [2]

21.3

1 a i 2.44 **ii** 2.94 **iii** 2.59 [3]
 b 4.88 [1]
2 a They are equal. [1]
 b [HA(aq)] is much greater than [H$^+$(aq)]. [1]

21.4

1 a A [1]
 b This is the alkaline region and the alkali is the same in both cases. [1]
 c i The equivalence point. **ii** Here the acid has been just neutralised by the alkali. [2]

21.5

1 There is no clear end point. There would be several gradual colour changes. [2]
2 Strong acid–strong base and strong acid–weak base. [2]

21.6

1 Sulfuric acid is a strong acid. [1]
2 a only. [1]
3 4.47 [1]

Answers to summary questions

22.1

1.   0   +1 −2   +2 −2   0
   $Mg(s) + H_2O(g) \rightarrow MgO(aq) + H_2(g)$
   Note that the sum of the oxidation states is zero on both sides of the arrow. Remember there are two atoms of hydrogen in the water molecule. [4]
2. P: $(+5) \times 2 = +10$, O: $5 \times (-2) = -10$. $+10 - 10 = 0$ [2]
3. Na: +1; O = −1 NB Peroxides are an exception to the rule that oxygen normally forms oxidation state −2. [2]

22.2

Go further

a. 84.7% (Atom economy is the mass of the required product, $CaCO_3$ in this case, divided by the total mass of the reactants expressed as a percentage.)

b. The limewater test for carbon dioxide. A solution of calcium hydroxide is called limewater. When carbon dioxide is bubbled into the clear solution, insoluble calcium carbonate is formed which turns the solution cloudy.

c. $CaCO_3(s) + H_2O(l) + CO_2(g) \rightarrow Ca(HCO_3)_2(aq)$

Summary questions

1. $H_2PO_4^-(aq) \rightleftharpoons H^+(aq) + HPO_4^{2-}(aq)$ [2]
   $HPO_4^{2-}(aq) \rightleftharpoons H^+(aq) + PO_4^{3-}(aq)$ [2]
2. +1 −2   +1 −2   +1 −2 +1
   $Na_2O(s) + H_2O(l) \rightarrow 2NaOH(aq)$
   This is not a redox reaction as no element changes its oxidation state. [3]
3. a. Burn phosphorus in a limited supply of oxygen. [1]
   b. $P_4O_6(s) + 6H_2O(l) \rightarrow 4H_3PO_3(aq)$ [2]

22.3

1. $SO_3(g) + 2KOH(aq) \rightarrow K_2SO_4(aq) + H_2O(l)$ [2]
2. +1 −2 +1   +1 +5 −2   +1 +5 −2 + 1 −2
   $3KOH(aq) + H_3PO_4(aq) \rightarrow K_3PO_4 + 3H_2O(l)$ [4]
3. $Na_2O(s) + H_2SO_4(aq) \rightarrow Na_2SO_4(aq) + H_2O(l)$ [2]
   $MgO(s) + 2HCl(aq) \rightarrow MgCl_2(aq) + H_2O(l)$ [2]

23.1

1. They all have their *outer* electrons in 4s. [1]
2. a. $1s^2\, 2s^2\, 2p^6\, 3s^2\, 3p^6\, 3d^2\, 4s^2$ [1]
   b. $[Ar]\, 3d^2\, 4s^2$ [1]
   c. $[Ar]\, 3d^2$ [1]
3. a. The two 4s electrons. [1]
   b. It has a half full 3d shell. [1]

23.2

Go further

a. Xenon is in period 5 and can use d-orbitals.
b. 12
c. An octahedron.

d. 2
e. Square planar.
f. Lone pairs are closer to the nucleus than shared pairs and repel more. They take up positions as far apart as possible.

Summary questions

1. a. $2 \rightarrow 7$ [1]
   b. More entities make the system more random. [1]
2. Carbon monoxide is a better ligand than oxygen and will replace oxygen carried by haemoglobin in the blood stream. [2]
3. a. Br^- [1]
   b. Fewer Br^- ions will fit round the metal so the complex is more likely to be tetrahedral than octagonal. [2]
4. a. The two Ns. [1]
   b. They have lone pairs of electrons. [1]
   c. Bidentate [1]

23.3

1. The Sc^{3+} ion has no part-filled d orbitals. [1]
2. a. Green [1]
   b. They are absorbed by the solution. [1]
3. The change in ligand, the change in co-ordination number (change of shape) and the charge (any two). [2]

23.4

1. 2:5 [1]
2. a. +7 −2   +1   +3 −2
   $2MnO_4^-(aq) + 16H^+(aq) + 5C_2O_4^{2-}(aq) \rightarrow$
    +4 −2   +2   +1 −2
   $10CO_2(g) + 2Mn^{2+}(aq) + 8H_2O(l)$ [1]
   b. Mn reduced, C oxidised. [1]
3. $Zn(s) + 2VO_2^+(aq) + 4H^+(aq) \rightarrow 2VO^{2+}(aq) + Zn^{2+}(aq) + 2H_2O(l)$ [1]

23.5

1. Multiply the top equation by 2 and then add the two half equations. [2]
2. It drops from 5 to 4 and then goes back to 5. [2]
3. a. $S_2O_8^{2-}(aq) + 2I^-(aq) \rightarrow 2SO_4^{2-}(aq) + I_2(aq)$ [2]
   b. i Reducing agent; ii oxidising agent. [2]
   c. Each of the two steps involves ions of opposite charge. The uncatalysed reaction involves two ions of the same charge. [2]

24.1

1. Because the high charge density of the M^{3+} ion attracts electrons towards it, weakening the O-H bonds in the water ligands and releasing H^+ ions. [1]
2. a. Bubbles of carbon dioxide gas. [1]
   b. The Fe^{3+} ion is sufficiently acidic to react with carbonate ions. [1]

203

Answers to summary questions

3 The bigger the value of K_a, the more H^+ ions dissociate, but pK_a is $-\log_{10} K_a$ so the bigger K_a gives a smaller pK_a. [2]

24.2/24.3

1 $[Cu(H_2O)_4(OH)_2](s) + 4NH_3(aq) \rightarrow$
$[Cu(NH_3)_4(H_2O)_2]^{2+}(aq) + 2H_2O(l) + 2OH^-(aq)$ [2]

2 Fe^{3+} is sufficiently acidic to react with the carbonate ion to form carbon dioxide gas. [1]

3 $1s^2\ 2s^2\ 2p^6$. It has no part-filled d orbitals to absorb light. [2]

25.1

1 a 2-fluorobutane [1]
  b 2-fluoro-2-methylbutane [1]
  c 1,2-difluorobutane [1]

2 a [2]
  b [2]
  c [2]

25.2

1 CHFClBr. This is the only compound with four different groups attached to the same carbon atom. [2]

2 [2]

3 $H_2C(CO_2)(NH_2)$. This is the only α-amino acid that does *not* have four different groups bonded to the central carbon atom. [2]

25.3

1 No, the central carbon atom of the product does not have four different groups bonded to it (there are two CH_3 groups). [2]

2 They have exactly the same relative molecular mass and degree of polarity in their bonds so van der Waals forces and dipole–dipole forces are exactly the same. They are the same shape so they pack together in the solid state in the same way. [2]

3 a [1]

b The two isomers might have different properties as drugs – if one is the active drug, the other may be inactive or toxic. [1]

26.1

1 Methanal as it has the lowest relative molecular mass and therefore the weakest van der Waals forces between its molecules, dipole–dipole forces remaining the same. [2]

2 A hydrogen bond requires a hydrogen bonded to an electronegative atom such as oxygen. Such hydrogens exist in water molecules but not in propanone. [2]

3 a [1]

$$CH_3CH_2\overset{O}{\underset{\|}{C}}CH_2CH_3$$

  b Pentanal. [1]

26.2

1 It has a negative charge and a lone pair of electrons with which to form a bond with $C^{\delta+}$. [1]

2 The nucleophile $:H^-$ will not attack the electron-rich C=C. [1]

3 [3]

26.3/26.4

Go further

a $RCOONa(aq) \rightleftharpoons RCOO^-(aq) + Na^+(aq)$

b It moves to the right forming an insoluble long-chain carboxylic acid which will not wash off the paper.

c The R-group of the acid is mostly non-polar hydrocarbon which will not mix with water. The salt is ionic and, like most ionic compounds, is water-soluble.

Summary questions

1 [1]

$$H_3C-\overset{O}{\underset{\|}{C}}-O-CH_2CH_3$$

2 3-bromopropanoic acid [1]

3 Propanoic acid and methanol [2]

4 a Ethanol and butanoic acid. **b** Ethanol and sodium butanoate. [4]

26.5

1

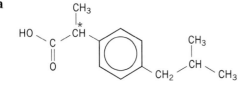

[2]

Answers to summary questions

2 Oxygen is more electronegative than nitrogen, so water donates its lone pair of electrons less readily than ammonia. [2]

3 75% [2]

27.1

1 $-240\,kJ\,mol^{-1}$ [1]

2 0.144 nm. (The measured value is 0.144 nm.) [1]

3 [1]

27.2/27.3

1 1,3-dichlorobenzene, [2]

2 a $C_6H_6 + NO_2^+ \rightarrow C_6H_5NO_2 + H^+$ [2]
  b $C_6H_5NO_2 + 3H_2 \rightarrow C_6H_5NH_2 + 2H_2O$ [2]

3 $RCO^+ + C_6H_6 \rightarrow C_6H_5COR + H^+$, the H^+ reacts with $AlCl_4^-$ to form $AlCl_3 + HCl$ [2]

4 A ketone. [1]

28.1

1 a Secondary. It has two organic groups bonded to the nitrogen atom. [2]
  b Methylphenylamine. [1]
  (Strictly, N-methylphenylamine to show that the methyl group is bonded to the nitrogen atom and is not a substituent on the benzene ring.)

2 [3]

28.2

1 a $(CH_3)_2NH_2Cl$ [1]
  b Ethylammonium chloride, $C_2H_5NH_3Cl$ [1]
  c Yes, because it is an ionic compound. [2]

2 a $(CH_3)_2NH$, $C_2H_5NH_2$. b Dimethylamine. It is a secondary amine and so has two inductive effects. [4]

28.3

Go further

a 4 CH$_2$—ONO$_2$
    |
    CH—ONO$_2$
    |
    CH$_2$—ONO$_2$ [1]

  $\rightarrow 12CO_2(g) + 6N_2(g) + 10H_2O(g) + O_2(g)$

b 29

c The production of 29 moles of gas from four moles of solid means that the reaction has a large positive entropy change, ΔS.

Summary questions

1 A mixture of primary, secondary, and tertiary amines will be formed along with a quaternary ammonium salt. These would have to be separated to produce a pure product. [2]

2 100%. It is an addition reaction so all the product is useful. [1]

3 A nitrile can have only one R group, so only primary amines can be produced. [1]

29.1

1 $HOCH_2CH_2OH$ and [2]

2 HCl [1]

3 H^+ and OH^-. [2]

30.1/30.2

1 Primary amine and carboxylic acid. [2]

2 It has a lone pair of electrons. [1]

3 a

  b [2]

4 [2]

5 A hydrogen bond forms between an O, N, or F atom and a hydrogen atom bonded to an O, N, or F atom. [2]

6 [2]

Answers to summary questions

30.3

1. If the oxygen atom uses its lone pair of electrons to accept a proton (H^+ ion), the lone pair will not be able to take part in hydrogen bonding. [2]
2. If an OH^- ion removes a H^+ ion from the COOH group, the hydrogen atom will not be available to form a hydrogen bond. [2]
3. Ionic bond; hydrogen bond; dipole–dipole bond; van der Waals forces. [2]

30.4

1. Phosphate, sugar, and base. [3]
2. TTGGCACAG [1]
3. The free OH group of the phosphate. [1]

30.5

1. A lone pair of electrons. [1]
2. Co-ordinate (dative) bonds. [1]
3. +2 [1]

31.1

1. a The more steps, the lower the overall yield.
   b The higher the atom economy the less waste. [2]
2. 1-chlorobutane → butan-1-ol → butananoic acid [2]
3. ethane → chloroethane → ethanol → ethanal [3]

31.2

1. a aromatic ring, carboxylic acid [2]
   b i Mass spectrometry [1]
   ii 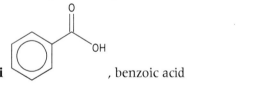 , benzoic acid [2]
   c 2-methylbenzoic acid, 3-methylbenzoic acid, 4-methylbenzoic acid [3]
   d i methyl benzoate ii ester [2]
   e They are isomers, so they will all have exactly the same M_r. [2]

32.1

1. a Compounds with the same formula but different arrangement of atoms in space. [1]
   b propanone   propanal [2]
   c i 2 ii 3 [2]
   d Figure 3 is propanone and Figure 4 is propanal. [1]
   e For propanone, δ = 200 is R**C**O (aldehyde or ketones); δ = 30 is RCO**C**
   For propanal, δ = 200 is R**C**O (aldehyde or ketones); δ = 36 is RCO**C**; δ = 10 is C–**C** [5 marks]

32.2

1. All the hydrogen atoms are in exactly the same environment. [1]
2. a 2 lines [1]   b 3:2 [1]
   c One line from the six hydrogens of the CH_3 groups and one from the four hydrogens on the benzene ring. [1]

32.3

1. a TMS. It is used to calibrate the spectrum because it appears at δ = 0. [2]
   b δ = 4.8 OH; δ = 3.5 CH_2; δ = 1.1 CH_3. [3]
   c The peak at δ = 4.8 is not split therefore there is no adjacent hydrogen.
   The peak at δ = 3.5 is split 1 : 3 : 3 : 1 so there are three adjacent hydrogens.
   The peak at δ = 1.1 is split 1 : 2 : 1, so there are two adjacent hydrogens. [3]
   d There are three different types of hydrogen atom but only two different types of carbon. [2]
2. None of the hydrogen atoms has a different type of hydrogen on an adjacent atom. [1]
3. The peaks at δ = 3.7 and 1.2 are methyl groups as each represents 3 hydrogen atoms. The one at 3.7 is not split, so it has no hydrogen atoms on an adjacent carbon. That at δ = 1.2 is split into three (1 : 2 : 1) so it has two hydrogen atoms on an adjacent carbon.
The peak at δ = 2.4 represents two hydrogen atoms. It is split into four (1 : 3 : 3 : 1) so it has three adjacent carbon atoms, ie a CH_2 group.
Using the table we can assign the peaks as follows: [6]

33.1

1. Top is 0.67, middle 0.50, bottom 0.17 [3]
2. The spot cannot travel further than the solvent. [1]
3. Amino acids are too similar in properties for most other techniques to be used. [1]

Answers to synoptic questions

1 a Chlorofluorocarbon [1]

[The halogen substituents are named in alphabetical order.]

b Aerosol propellants, foam blowing, degreasing solvents, fire extinguishers. Any two of these. [2]

c Ozone absorbs ultraviolet radiation which can cause skin cancers, cause photochemical smogs, and damage paint and plastics. Any two of these. [2]

d i They have an unpaired electron. [1]

ii Free radicals (or just 'radicals'). [1]

iii $O_3 + O \rightarrow 2O_2$ [2]

[The overall reaction is obtained by adding reactions 1 and 2 and cancelling species that appear on both sides of the arrow.]

$Cl\bullet + O_3 \rightarrow ClO\bullet + O_2$

$ClO\bullet + O \rightarrow O_2 + Cl\bullet$

~~Cl•~~ + O_3 + ~~ClO•~~ + O → ~~ClO•~~ + O_2 + O_2 + ~~Cl•~~

iv It is a catalyst because it is not used up in the reaction. [2]

e i $Cl\bullet + Cl\bullet \rightarrow Cl_2$ [1]

[In a termination step, free radicals are destroyed.]

ii The concentration of Cl· in the atmosphere is very small. [1]

f i Rate = k [Cl•] [O_3] [1]

ii 2 [1]

[The overall order is the sum of the orders with respect to each species.]

g There was a great deal of CFCs in the atmosphere. Each Cl• can destroy many molecules of ozone before a termination step occurs. There is a large reservoir of CFCs 'locked up' in old fridges, aerosol cans, etc. which is gradually being released. It takes time for CFCs at ground level to reach the upper atmosphere. Any two of these. [2]

h It has a higher M_r and therefore more electrons. So the van der Waals forces between the molecules will be greater. [2]

i The C–I bond is weaker than C–F, i.e. the C–I bond has a lower bond enthalpy. [1]

j Add silver nitrate solution, and a pale yellow precipitate will form that is insoluble in concentrated ammonia solution. The reaction mixture must be neutralised by adding nitric acid to remove OH⁻ ions which would also form a precipitate with silver nitrate. [5]

[Neither HCl nor H_2SO_4 can be used because they would form a precipitate with silver nitrate.]

2 a A: 2-bromobutane; B: but-1-ene; C: butan-2-ol [3]

b i Reflux with potassium hydroxide (or sodium hydroxide) dissolved in ethanol. This is an elimination reaction. [4]

ii React with cold, aqueous potassium hydroxide (or sodium hydroxide). This is a (nucleophilic) substitution reaction. [4]

[Notice that the reagent is the same, but the conditions differ.]

c i $CH_3CH=CH_2H_3$, but-2-ene. [2]

ii E-Z (or cis-trans) isomerism. [1]

iii

, E-but-2-ene (or trans-but-2-ene)

, Z-but-2-ene (or cis but 2-ene) [4]

[Displayed formulae show every atom and every bond.]

d React with C≡N⁻ ions and then with HCl. [2]

[Adding one carbon atom always suggests the use of C≡N⁻ ions.]

e

[1]

Optical isomerism occurs when there is a carbon atom that is bonded to four different groups. [2]

3 a The two nitrogen oxides [collectively called NO_x] react in the air to form nitric acid which causes acid rain. They also react with unburnt hydrocarbons to form photochemical smog. Carbon monoxide is a poisonous gas. [3]

b This is produced by incomplete combustion of the hydrocarbon fuel. [1]

c i $N_2(g) + O_2(g) \rightarrow 2NO(g)$ [2]

ii $N_2(g) + 2O_2(g) \rightarrow 2NO_2(g)$ [2]

[Always check equations for balance – they should have the same number of atoms of each element on each side of the arrow.]

d Carbon dioxide is not poisonous, although it is a greenhouse gas. Nitrogen is unreactive. [2]

e i 0, **ii** +2, **iii** +4. [3]

[Remember oxygen (almost) always has an oxidation state of –2 in its compounds, and the sum of the oxidation states in a neutral compound is zero. The oxidation state of an uncombined element is always 0.]

Answers to synoptic questions

f The N≡N triple bond is very strong and will only break at high temperatures. [2]

[The bond enthalpy is **945 kJ mol⁻¹ compared with 347 kJ mol⁻¹ for C–C.**]

g Heterogeneous. This means that the catalyst is in a different phase from the reactants. **[Here solid catalyst, gaseous reactants.]** Homogeneous means that the catalyst is in the same phase as the reactants. [3]

h It lowers the activation energy of the reaction. [1]

i To increase its surface area. [1]

[**In heterogeneous catalysis the reaction occurs on the catalyst surface.**]

j

[**Start with the displayed formula if you are unfamiliar with the skeletal form.**] [1]

k Lead accumulates in the organs and poisons the body. It also accumulates on the catalyst and 'poisons' it. **[Either answer is acceptable.]** [1]

l Condensed water formed from the hydrogen of the hydrocarbon fuel combining with oxygen from the air. [2]

4 a A mole of lithium hydroxide is lighter than a mole of sodium hydroxide and weight is important in spacecraft. **[Both absorb the same amount of carbon dioxide.]** [2]

b $2NaOH(s) + CO_2(g) \rightarrow Na_2CO_3(s) + H_2O(l)$ [1]

c i 20.8 mol **ii** 41.6 mol **iii** 994.2 g **[The mass of 1 mole of LiOH is 23.9 g.] iv** 1663.9 **[The mass of 1 mole of NaOH is 40 g.]** [4]

d i Because it is a gas. [1]

   ii −160.4 J K⁻¹ mol⁻¹; negative because a gas (random) is turning into a solid (more ordered). [2]

[**ΔS is entropy of products − entropy of reactants**]

   iii −178.3 kJ mol⁻¹ [1]

   iv $\Delta G = \Delta H - T\Delta S = -178.3 - (298 \times -160.4 / 1000) = -130.5$ kJ mol⁻¹ [2]

[**Remember to convert the entropy units to kJ K⁻¹ mol⁻¹ by dividing by 1000.**]

   v Yes, because ΔG is negative. [2]

   vi $\Delta G = \Delta H - T\Delta S = -178.3 - (1500 \times -160.4 / 1000) = +62.3$ kJ mol⁻¹; the reaction is not feasible **[because ΔG is positive]**. [3]

   vii Nothing. [1]

[**The sign of ΔG tells us that the reaction is feasible but nothing about the rate.**]

5 a **Optical isomerism** – existing as a pair of non-superimposable mirror image isomers. [2]
Chiral centre – a carbon atom bonded to four different groups. [2]
Racemate – a 50 : 50 mixture of the two optical isomers. [2]

b [1]

[**The chiral carbon is bonded to four different groups. It may help to draw in the hydrogen atoms which are not shown on this skeletal formula.**]

c They rotate the plane of polarisation of polarised light in opposite directions. [2]

d Racemates are difficult/expensive to separate. The other isomer is not harmful. [2]

e It is mostly non-polar – only the COOH group will form hydrogen bonds with water. [1]

f i An amino acid with the COOH and NH₂ on the same carbon atom. [2]

   ii It has three groups that can form hydrogen bonds with water, 2 × NH₂ and 1 × COOH. [1]

   iii It is present in the body. [1]

g One of the basic NH₂ groups could react with the acidic COOH group of ibuprofen to form a salt. [1]

h i 200 × 334 / 206 = 324.3 mg [2]

   ii A molecule of water (M_r 18) is eliminated in the reaction. [1]

[**The two form a salt with water as a by-product.**]

6 a i 0.2 mol CO_2 **ii** 0.3 mol H_2O [2]

[**To find the number of moles, divide the mass in grams by A_r.**]

b i 0.2 mol C [1]   **ii** 2.4 g C [1]

c 0.6 mol (ii) 0.6 g [2]

d i 3.0 g [1]   **ii** 1.6 g [1]   **iii** 0.1 mol [1]

e C_2H_6O [1]

f i Mass spectrometry [1]

   ii C_2H_6O [1]

[**M_r of the empirical formula is 46, so the empirical and molecular formulae are the same.**]

g i This is an O–H stretch so A has an O–H group [2]

   ii It must be CH_3CH_2OH. [1]

   iii ethanol, alcohols [1]

h Warm with yellow acidified potassium dichromate solution. The solution would turn to green. [2]

i They are isomers and would therefore give exactly the same result. [2]

j i 3;   **ii** 1 [2]

[**All the hydrogen atoms in CH_3OCH_3 are in identical environments.**]

k i 2   **ii** 1 [1]

[**Both carbon atoms in CH_3OCH_3 are in identical environments.**]

Answers to synoptic questions

7 a $CaCO_3(s) + 2HCl(aq) \rightarrow CaCl_2(aq) + H_2O(l) + CO_2(g)$ *[2]*

  b i 36.5 *[1]* **ii** 0.27 mol *[1]* **iii** 2.7 mol dm^{-3} *[1]*

  c i $HCl(aq) + NaOH(aq) \rightarrow NaCl(aq) + H_2O(l)$ *[1]*

    ii Colourless to pink. *[1]*

    iii The colour change is more distinct than red to yellow. *[1]*

  d i 0.07 mol *[1]*

[Use the equation No. of moles = c × V / 1000.]

    ii 0.07 mol *[1]*

    iii 2.8 mol dm^{-3} *[1]*

    iv Yes *[1]*

  e i   +1+1−2   +1−1   0   +1−2
     $HClO(aq) + HCl(aq) \rightarrow Cl_2(g) + H_2O(l)$ *[2]*

    ii Chlorine is a toxic gas. *[1]*

    iii 50 cm^3 Kleeno contains 50 × 2.8 / 1000 mol = 0.14 mol. This produces 0.14 × 24 dm^3 of gas at room conditions. 3.4 dm^3 chlorine. **['Excess' means that there is enough bleach to react with all the Kleeno.]** *[2]*

8 a $Mg^+(g) \rightarrow Mg^{2+}(g) + e^-$ *[2]*

[The state symbols are essential. This is *not* $Mg(g) \rightarrow Mg^{2+}(g) + 2e^-$]

  b As each successive electron is removed, it comes from a more highly positively charged ion. *[2]*

  c There is a large jump in IE after two electrons have been lost and another jump after a further eight electrons have been lost. *[2]*

  d Magnesium has three isotopes of mass numbers 24, 25, and 26. *[2]*

  e $\dfrac{(24 \times 78.7) + (25 \times 10.13) + (26 \times 11.17)}{100} = 24.32$ *[2]*

  f 12 protons, 13 neutrons, 12 electrons. *[3]*

  g i $Mg(s) + 2H_2O(l) \rightarrow Mg(OH)_2(s) + H_2(g)$ *[2]*
      0   +1−2   +2 −2+1   0

    ii $Mg(s) + 2H_2O(l) \rightarrow Mg(OH)_2(s) + H_2(g)$ *[2]*

[Remember the oxidation state of an uncombined element is always 0. The sum of the oxidation states in a neutral compound is zero.]

  h Giant ionic. *[2]*

  i The solubility increases. *[1]*

Data

Constants

Gas constant

$R = 8.31 \, \text{J K}^{-1} \text{mol}^{-1}$

Specific heat capacity

$c = 4.2 \, \text{J g K}^{-1}$

Avagadro's constant

6.022×10^{23}

Infrared

Bond	Wavenumber / cm^{-1}
C—H	2850–3300
C—C	750–1100
C=C	1620–1680
C=O	1680–1750
C—O	1000–1300
O—H (alcohols)	3230–3550
O—H (acids)	2500–3000
N—H	3300–3500

Table A Infrared

Bond	Wavenumber / cm^{-1}
C—H	2850–3300
C—C	750–1100
C=C	1620–1680
C=O	1680–1750
C—O	1000–1300
O—H (alcohols)	3230–3550
O—H (acids)	2500–3000
N—H	3300–3500

Table B ^{1}H NMR chemical shift data

Type of proton	δ/ppm
ROH	0.5–5.0
RCH$_3$	0.7–1.2
RNH$_2$	1.0–4.5
R$_2$CH$_2$	1.2–1.4
R$_3$CH	1.4–1.6
R—C(=O)—C—H	2.1–2.6
R—O—C—H	3.1–3.9
RCH$_2$Cl or Br	3.1–4.2
R—C(=O)—O—C—H	3.7–4.1
R$_2$C=CHR	4.5–6.0
R—C(=O)H	9.0–10.0
R—C(=O)O—H	10.0–12.0

Table C ^{13}C NMR chemical shift data

Type of carbon	δ/ppm
—C—C—	5–40
R—C—Cl or Br	10–70
R—C—C(=O)—	20–50
R—C—N	25–60
—C—O— alcohols, ethers or esters	50–90
C=C	90–150
R—C≡N	110–125
benzene ring	110–160
R—C(=O)— esters or acids	160–185
R—C(=O)— aldehydes or ketones	190–220

Phosphate and sugars

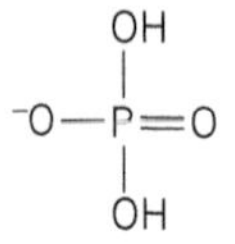

phosphate

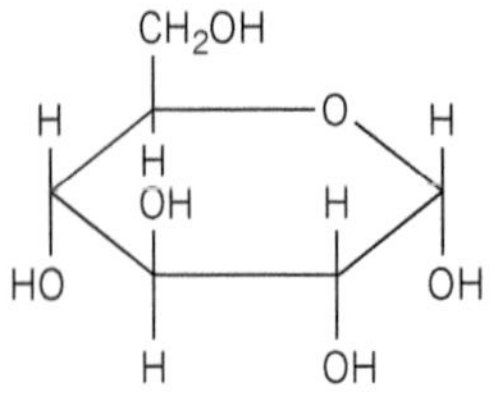

glucose

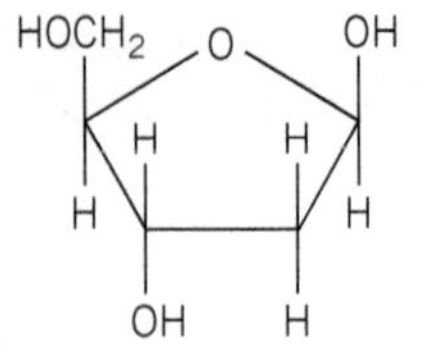

2-deoxyribose

Bases

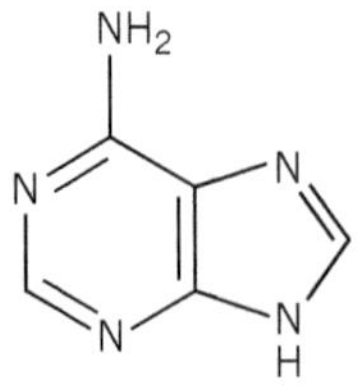

adenine

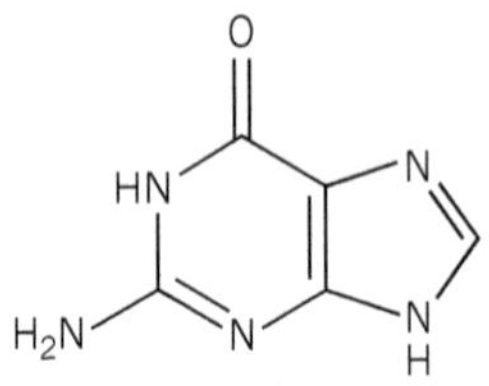

guanine

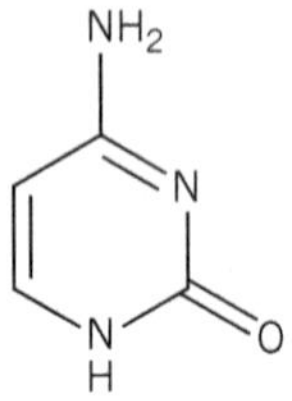

cytosine

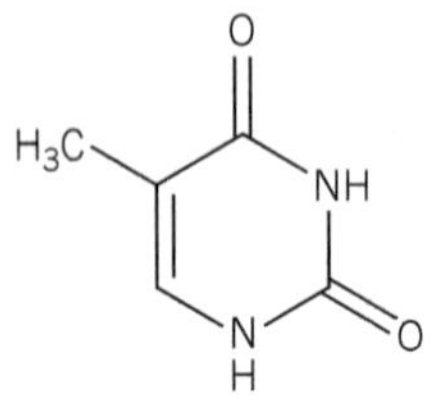

thymine

Amino acids

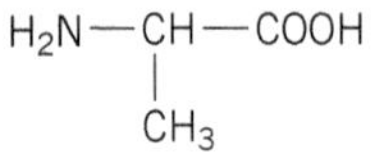

alanine

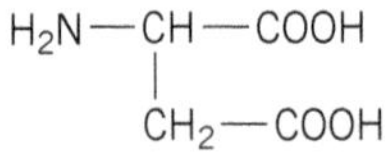

aspartic acid

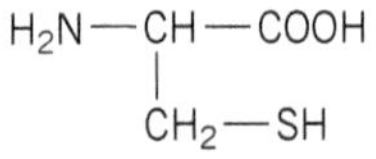

cysteine

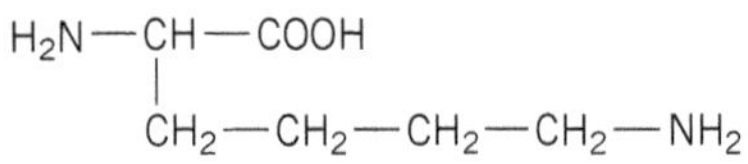

lysine

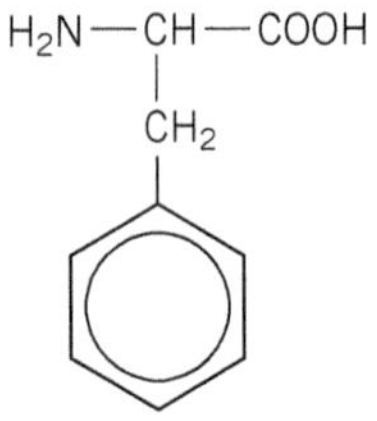

phenylalanine

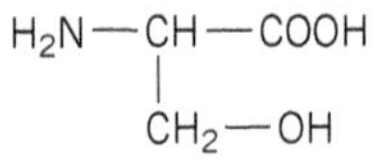

serine

Haem B

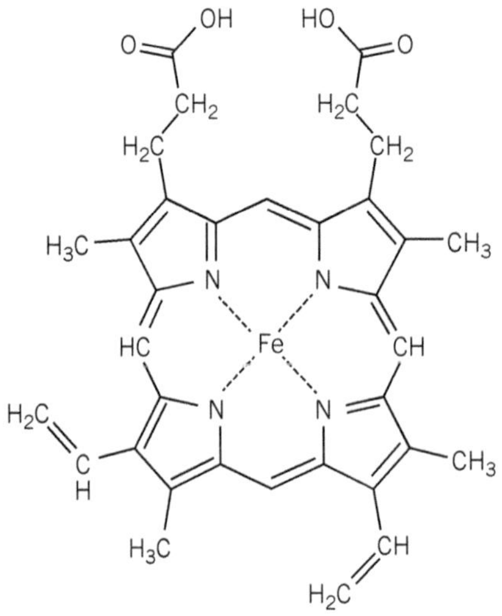

Periodic Table

The Periodic Table of the elements

Key

relative atomic mass
atomic symbol
name
atomic (proton) number

(1)	(2)		(3)	(4)	(5)	(6)	(7)	(8)	(9)	(10)	(11)	(12)	(13)	(14)	(15)	(16)	(17)	(0)
1	2												3	4	5	6	7	(18)
								1.0 **H** hydrogen 1										4.0 **He** helium 2
6.9 **Li** lithium 3	9.0 **Be** beryllium 4												10.8 **B** boron 5	12.0 **C** carbon 6	14.0 **N** nitrogen 7	16.0 **O** oxygen 8	19.0 **F** fluorine 9	20.2 **Ne** neon 10
23.0 **Na** sodium 11	24.3 **Mg** magnesium 12												27.0 **Al** aluminium 13	28.1 **Si** silicon 14	31.0 **P** phosphorus 15	32.1 **S** sulfur 16	35.5 **Cl** chlorine 17	39.9 **Ar** argon 18
39.1 **K** potassium 19	40.1 **Ca** calcium 20		45.0 **Sc** scandium 21	47.9 **Ti** titanium 22	50.9 **V** vanadium 23	52.0 **Cr** chromium 24	54.9 **Mn** manganese 25	55.8 **Fe** iron 26	58.9 **Co** cobalt 27	58.7 **Ni** nickel 28	63.5 **Cu** copper 29	65.4 **Zn** zinc 30	69.7 **Ga** gallium 31	72.6 **Ge** germanium 32	74.9 **As** arsenic 33	79.0 **Se** selenium 34	79.9 **Br** bromine 35	83.8 **Kr** krypton 36
85.5 **Rb** rubidium 37	87.6 **Sr** strontium 38		88.9 **Y** yttrium 39	91.2 **Zr** zirconium 40	92.9 **Nb** niobium 41	95.9 **Mo** molybdenum 42	[98] **Tc** technetium 43	101.1 **Ru** ruthenium 44	102.9 **Rh** rhodium 45	106.4 **Pd** palladium 46	107.9 **Ag** silver 47	112.4 **Cd** cadmium 48	114.8 **In** indium 49	118.7 **Sn** tin 50	121.8 **Sb** antimony 51	127.6 **Te** tellurium 52	126.9 **I** iodine 53	131.3 **Xe** xenon 54
132.9 **Cs** caesium 55	137.3 **Ba** barium 56		138.9 **La*** lanthanum 57	178.5 **Hf** hafnium 72	180.9 **Ta** tantalum 73	183.8 **W** tungsten 74	186.2 **Re** rhenium 75	190.2 **Os** osmium 76	192.2 **Ir** iridium 77	195.1 **Pt** platinum 78	197.0 **Au** gold 79	200.6 **Hg** mercury 80	204.4 **Tl** thallium 81	207.2 **Pb** lead 82	209.0 **Bi** bismuth 83	[209] **Po** polonium 84	[210] **At** astatine 85	[222] **Rn** radon 86
[223] **Fr** francium 87	[226] **Ra** radium 88		[227] **Ac†** actinium 89	[261] **Rf** rutherfordium 104	[262] **Db** dubnium 105	[266] **Sg** seaborgium 106	[264] **Bh** bohrium 107	[277] **Hs** hassium 108	[268] **Mt** meitnerium 109	[271] **Ds** darmstadtium 110	[272] **Rg** roentgenium 111							

Elements with atomic numbers 112-116 have been reported but not fully authenticated

* 58 – 71 Lanthanides

† 90 – 103 Actinides

140.1 **Ce** cerium 58	140.9 **Pr** praseodymium 59	144.2 **Nd** neodymium 60	144.9 **Pm** promethium 61	150.4 **Sm** samarium 62	152.0 **Eu** europium 63	157.3 **Gd** gadolinium 64	158.9 **Tb** terbium 65	162.5 **Dy** dysprosium 66	164.9 **Ho** holmium 67	167.3 **Er** erbium 68	168.9 **Tm** thulium 69	173.0 **Yb** ytterbium 70	175.0 **Lu** lutetium 71
232.0 **Th** thorium 90	231.0 **Pa** protactinium 91	238.0 **U** uranium 92	237.0 **Np** neptunium 93	239.1 **Pu** plutonium 94	243.1 **Am** americium 95	247.1 **Cm** curium 96	247.1 **Bk** berkelium 97	252.1 **Cf** californium 98	[252] **Es** einsteinium 99	[257] **Fm** fermium 100	[258] **Md** mendelevium 101	[259] **No** nobelium 102	[260] **Lr** lawrencium 103